Buying and Setting Up Your Small Farm or Ranch

L. R. Miller

Buying and Setting Up Your Small Farm or Ranch
Copyright © 1997 Lynn R. Miller

All rights reserved, including those of translation. This book, or parts thereof, may not be reproduced in any form without the written permission of the author or publisher. Neither the author nor the publisher, by publication of this material, ensure to anyone the use of such material against liability of any kind including infringement of any patent. Inquiries should be addressed to Small Farmer's Journal, Inc. PO Box 1627, Sisters, Oregon 97759.

Publisher
Small Farmer's Journal Inc.
PO Box 1627
325 Barclay Drive
Sisters, Oregon 97759
541-549-2064

Printed in the United States of America
Printer: Parton Press, Redmond, Oregon

Authored by Lynn R. Miller

First Edition,
First printing 1997

ISBN 1-885210-06-X

Also by L. R. Miller

The Work Horse Handbook

Training Workhorses / Training Teamsters

Why Farm: Essays & Editorials

Ten Acres Enough: The Small Farm Dream Is Possible

Thought Small: Poems, Prayers, Drawings & Postings

This book is dedicated to every person who ever held a dream for a small farm and searched for the possibility.

Buying and Setting Up Your Small Farm or Ranch

table of contents

Preface 8

Chapter One 13
Why Do You Want To Farm ... There?

Chapter Two 22
How Much Is That Farm Worth To You?

Chapter Three 35
Measuring and Researching

Chapter Four 50
Money Changing Hands

Chapter Five 58
After The Farm Is Bought

Chapter Six	67
Designing Your Farm	
Chapter Seven	83
Great Silly Hopes	
Chapter Eight	91
Costs of Starting A Horse-powered Farm	
Chapter Nine	108
Setting Up a Horse-powered Farm	
Chapter Ten	143
Setting Up a Tractor-powered Farm	
Chapter Eleven	154
Outbuildings / Sheds & Barns	
Chapter Twelve	175
Naming Your Farm	
Chapter Thirteen	182
For The Long Haul	
Index	192

"I write more particularly for those who have not been brought up as farmers- for that numerous body of patient toilers in city, town and village, who, like myself, have struggled on from year to year, anxious to break away from the bondage of the desk, the counter, or the workshop, to realize in the country even a moderate income, so that it be a sure one. Many such are constantly looking around in this direction for something which, with less mental toil and anxiety, will provide a maintenance for a growing family, and afford a refuge for advancing age- some safe and quiet haven, sheltered from the constantly recurring monetary and political convulsions which in this country so suddenly reduce men to poverty. But these inquirers find no experienced pioneers to lead the way, and they turn back upon themselves, too fearful to go forward alone."

*- Anonymous, from "**Ten Acres Enough**: A Practical Experience Showing How a Very Small Farm May be Made to Keep a Very Large Family with Extensive and Profitable Experience."*
1864 and 1996

preface

This book is born of a noble and important cause, we who edit and publish the international quarterly **_Small Farmer's Journal_** wish to assist people with honest and pertinent information on designing, acquiring and establishing a working small independent farm holding. This is done because, contrary to prevalent commercial notions about the human condition which describes us as blissful indolent pill-popping corporate citizens, mankind and the planet are still best served by the social model constructed from a fabric of small independent farms, businesses, and communities interwoven into a social blanket of magnificent diversity and great natural strength.

You want to be on a farm. We want to help get you there. You think a farm might be a better place to raise a family. We know it is. You think you would be happier farming because it looks like it will make you feel better about yourself. That is for sure. We'll go even further, the world needs more small farmers like yourself as quick as it can get them. Mankind and the planet are at grave risk if the modern trends towards big industrial farming complexes continues. Small Farms, small towns, and communities built on the nucleus of families are the solution.

Most everywhere you go today people will volunteer that your dream of a small farm of your own is hopeless. They don't know this for a fact, they assume its true because they hear it and because perhaps they knew folks who were driven out by the banking frauds of the eighties. But if you go to the right corners you'll stumble onto folks who are, today, doing just exactly what you want to do. They are doing it well and they'll be the first to tell you that your dream is hopeful and possible. But you won't be able to get there without some information and a little guidance through the maze of procedures for buying and setting up a farm. This book is one small effort to provide you with that help.

Don't accept what you read here as gospel, look for additional sources of information and be in charge of the design of your farming adventure from the

"...mankind and the planet are still best served by the social model constructed from a fabric of small independent farms, businesses, and communities interwoven into a social blanket of magnificent diversity and great natural strength."

very beginning. You may already have a feeling that you need to be farming, and that is fine. But lest you come to wonder if your dream is selfish let me tell you that the world needs small independent farmers who work every day with craftmanship and genuine concern for all they touch and create. By becoming what you want to be you also serve mankind and the planet. Your personal dream, when it becomes reality, will lead you into a life of good service and great reward. You are in good company and you are good company. All my best wishes for your journey.

- *L. R. Miller, Singing Horse Ranch, 1997*

Who was that early sodbuster in Kansas? He leaned at the gatepost and studied the horizon and figured what corn might do next year and tried to calculate why God ever made the grasshopper and why two days of hot winds smother the life out of a stand of wheat and why there was such a spread between what he got for grain and the price quoted in Chicago and New York. Drove up a newcomer in a covered wagon: "What kind of folks live around here?" "Well stranger, what kind of folks was there in the country you come from?" "Well, they was mostly a lowdown, lying, thieving, gossiping, backbiting lot of people." "Well, I guess, stranger, that's about the kind of folks you'll find around here." And the dusty gray stranger had just about blended into the dusty gray cottonwoods in a clump on the horizon when another newcomer drove up: "What kind of folks live around here?" Well stranger, what kind of folks was there in the country you come from?" "Well, they was mostly a decent, hard-working, law-abiding, friendly lot of people." "Well, I guess stranger, that's about the kind of folks you'll find around here." And the second wagon moved off and blended with the dusty gray cottonwoods on the horizon while the early sodbuster leaned at his gatepost and tried to figure why two days of hot winds smother the life out of a nice stand of wheat.

- from THE PEOPLE, YES (1936) by Carl Sandburg

Chapter One

Why Do You Want To Farm...

There?

*"I would rather sit on a pumpkin, and have it all to myself,
than to be crowded on a velvet cushion."*
-Henry David Thoreau

What kind of farm do you want and where?

This book in hand, you are most likely thinking about acquiring a farm and looking for helpful information. You may not know you hold the answers to many of your questions. And many of the remaining queries cannot be answered without clear information from you. So we must begin by asking you questions and asking that you hold your answers to those questions uppermost in your mind as you work to select from a myriad of options.

We must first look at your motivations, or, simply put, WHY you want a farm. Our many years of work with **Small Farmer's Journal** has taught us, repeatedly, that valuable pertinent working answers to important questions come best when we evaluate or re-evaluate our motivations. Many of us benefit from having our motivations regularly questioned and re-evaluated.

We cannot assume that all who wish to acquire a farm property also desire to operate a farm. But you, as the reader of this text, should know we couch most of our presentation with the thought that you are wanting to make your new farm "work" by your design. For this reason many of the suggested preliminary considerations are vitally important.

Although complex, the more common procedures for actually buying a farm are less difficult than the necessary work on the initial design and the subsequent establishment of your farm operation. Selecting a farm, investigating that farm, getting financially prepared for the purchase - and, after purchase, modifying and implementing your farm operation design are all key elements critical to your success and requiring great foresight and planning. You might zip right through a farm purchase only to find out you bought the wrong place in the wrong way and cannot possibly realize your dream there.

Wanting a farm for the sake of your children: what more powerful reason could there be?.

FIRST THINGS FIRST:
WHY DO YOU WANT TO FARM?
or
WHY DO YOU WANT A FARM?
WHAT KIND OF FARM DO YOU WANT?
WHERE DO YOU WANT TO FARM
and
IS THAT THE BEST PLACE FOR YOUR DREAM?

WHY:

This is the single most important question and the one you SHOULD answer first. Perhaps a series of questions will help you to formulate your answer and see the many reasons why that it is so important.

ARE YOU EXPERIENCED WITH FARM WORK?

ARE YOU EDUCATED, BY EXPERIENCE OR SCHOOL OR BOTH, IN FARM MATTERS?

*WHY DO YOU WANT **A** FARM?*
- *a.) Is it for the income opportunities?*
- *b.) Is it for long-term investment?*
- *c.) Is it for the lifestyle?*
- *d.) Is it because it represents some sort of escape for you?*
- *e.) Do questions of health and welfare of your family figure in?*
- *f.) Is it for some other reason?*

*WHY DO YOU WANT **TO** FARM?*
- *a.) Because your father did?*
- *b.) Because it looks like a good life?*
- *c.) Because you're being forced to?*
- *d.) Because you feel you can make good money?*
- *e.) Because you feel called to this work?*
- *f.) Because it looks like the only sensible way to get out of where you are?*

Please believe there is no implied prioritization or value judgement made to go with any of the above questions. We are not suggesting that one reason is more valid than another, ONLY that the answers to such questions will help you

Do you want to farm because the experiences of the old home place keep calling you?

FIND what you're really looking for. The answer to any of these questions will point you in certain, maybe several, directions.

For example: If money's most important you'll want to be where the markets and population growth are happening (in a word, near lots of people). If you want to farm to enjoy the life-style the opposite may be true making a remote setting more attractive. If you're inexperienced you may need to put a premium on settling into a strong community of small farmers who share your philosophy; they'll be the best help for the newcomer.

It might work for some of you to write up a check list against which you can measure each region and each farm property you consider. Customize the list to reflect your own priorities. Here's one example of a list:

_____*do we really like it here - weather, look of the landscape and farming opportunities?*

_____*can a farm make a good living here?*

_____*would this farmland in this area be a good long term investment?*

_____*can my family and I be happy here?*

_____*are the neighboring folks the sort of community we want to belong to?*

_____*will this community accept us?*

In this way your personal motivations, needs and values can be used as a helpful measuring stick in the midst of that confusing first flush of excitement over a potential new farm home.

All that's necessary in the beginning is for you to have a clear fix on your motivation and goals. Understand why you want a farm and why you want to farm. Even if you find your answers changing on down the road, having a clear view of some answers now will make the process gel more quickly for you.

WHAT:

You'd be surprised how many people set out to buy a farm without any clear idea of what kind of farming they mean to get into.

You want to raise hogs?...milk cows?...grow vegetables?... tree fruit? ...raise poultry?... horses?... beef cattle?... sheep?... exotic livestock? ...grow catfish?... wheat?... mint?... cane fruit?... or any combination of the above?

Here's where we must return to the philosophical roots of *Small Farmer's Journal* and put in a time-worn plug:

DON'T PUT ALL YOUR EGGS IN ONE BASKET.

MIXED CROP AND LIVESTOCK OPERATIONS spread the risks and profit from interrelationships. (But more on that later.)

What you want to raise may limit the areas suitable for your farming, it may also point to areas where markets are better suited. Certain fruit and stock, like Loquats and Catfish, need to be in warmer climates whereas the cost of distilling Mint will probably point you in the direction of other Mint growers where equipment can be co-operatively set up.

If you have a strong idea of what combination you want to raise but little or no experience with same you will have to do some research to understand what the inherent limitations are of that production. Otherwise you could find yourself on a farm that won't raise blueberries, sheep and carrots profitably even though you have your heart set on it.

Most people have clear ideas of what kind of community they'd like to live near and what sort of climate they want. But it's surprising how few give consideration to the ways those two elements affect the suitability of their proposed operation.

WHERE:

Now, just when it may seem like this discussion is sorting out the concerns and answers into neat little piles I have to throw a wrench into the mechanism. The best possible place to raise certain crops, if you measure soils, weather (including precipitation), and apparent nearby markets, may seem like the obvious FIRST choice, however, for you it may not be the BEST choice. There are three VERY IMPORTANT offsetting factors.

1.) the net effect of the immediate farm community.

2.) the necessary adjustments for quality and hazard.

3.) your personal preferences.

You may find it easy to imagine how the right community of farming neighbors would make a farm more attractive. But consider that the opposite could be true and for unlikely reasons.

I once owned a farm that was in a major agribusiness valley. My cropland was separated from the neighboring fields by nothing more than a fence. Planes and helicopters swooped down on those fences and spread chemicals across neighboring fields while making my job of plowing with six head of workhorses an exciting challenge as the animals struggled to understand the aerobatics, the hissing sprays and the clouds of odd smelling chemicals. Across the street, people in the subdivision complained that our six hogs, three cows, and half dozen work horses were creating a health and noise hazard. Yet I felt that this farm, although not to my personal liking, represented the most sensible investment because of its proximity to markets, excellent soils, good water and mild climate. My industrialized farming neighbors treated our little horse-powered farm as though it were a freak show. Being interested in the use of less agricultural chemicals we got into discussion with organic FARMERS in

> *...the success of your farming operation will be jeopardized if you don't like where you are.*

the area and joined in the formation of organizations which, ostensibly, were supposed to be mutually beneficial to establishing market dialogs and cooperative farm input purchases. To my dismay what quickly happened was the establishment of an exclusive club that worked to corner markets and prevent other farmers access, regardless their methodology. It became a kind of unionization of organic farming that took much of the independence away from its members.

Add to this that although the soil and climate produced good-looking vegetables, fruit, hay and grains - the measured food value (and resultant flavor) were lacking. 70 inches of rain at the wrong times and the character of most of the farming in the region, along with the highly competitive and selfish nature of the community more than offset what were initially good points.

Compare that with where we are now. We get only 16 inches of rain and need to practice difficult and sensitive dry-farming along with adding expensive irrigation water. Our soils are rocky and sandy, highly susceptible to wind erosion. We are prone to frost damage. We are forced to deal with the logic (A=B if they say it does) of the Forest Service as they maintain the lands which border us. We have no neighbors for ten miles. The nearest town is sixteen miles away and has a population of 700. BUT - we love where we are every minute of every

"You may find it easy to imagine how the right community of farming neighbors would make a farm more attractive."

day. AND - we are forced to be better farmers because the rewards come harder. AND - we have a neighborhood that extends for a forty mile circle making the sharing of work and pleasure all the more special because it cannot be so lightly taken. AND - we've discovered that trying to farm here, where most others would not, gives us a delightful edge on local markets. (99 % of all produce is shipped into the area!) Add to this the fact that we don't have to deal with the over-the-fence farming practices of other farmers who choose to use heavy chemicals. Our horse-working practices go unfettered. And we enjoy the bonus of much higher food value and flavor to the produce and meats we grow. Not to forget that this place cost a tenth of what a comparable place would sell for in more popular and populated areas.

Is tranquility, security and a sense of self worth important to you and your family?

(Specific to our acquisition price were these limitations: no phone, no drinkable house water, limited winter access on unimproved dirt roads, no mail delivery, and heavy predator populations.)

I am not trying to sell anyone on the idea that they should consider this area we're in. It is probably only suitable for someone capable of creative rationale. I am using this example to point out that the BEST place may be so for uncommon reasons. And the most uncommon, and some might say impractical, reason is your personal preference. I strongly believe that the success of your farming operation will be jeopardized if you don't like where you are.

If you're anxious about buying a farm I can imagine how this discussion might frustrate you because you want to get on with how much you should pay, how to get financing, whether or not you need title insurance, how to "handle" real estate sales people, and those sorts of questions. But you've got to get through the front door in this process. You can't just stand outside and expect to get, by anxious osmosis, into that front room. You shouldn't break a window and climb in, although that's the way some do it. No, you have to find the front door, knock on it and be let in. After having bought three farms (sold one and lost one), and having been party to lots of discussions through the ***Small Farmer's Journal*** I feel compelled to try to share a suggestion of some flexible order to this important process. So, figure out why, what and where before you look at real estate and definitely before you sign anything!

Be where you want to be and you'll be there for the long haul.

Chapter Two

"We can lie to ourselves about many things; but if we lie about our relationship to the land, the land will suffer, and soon we and all other creatures that share the land will suffer. If we persist in our ignorance or dishonesty, we will die, as surely as those bighorns perish from not knowing where they are. We are smarter than sheep, in most respects. Seeing the danger of ignorance, we may be moved to invent or recover some of the lore that connects us to the land, and tells us how to live in our place."
- Scott Russell Sanders

Who are you?

And how much is that farm worth to you?

Last chapter we started a discussion on the self analyzing procedures that might go into considerations of the purchase of a farm. We discussed the important first questions including;

*Why you want to farm,
what kind of farm you want,
and where you want to farm?*

This chapter we'll look at how much you might, or should, pay for a farm.

Up until the late 1980's conventional farm lenders used to be proud of the formulas and yardsticks they employed to determine the value of a given piece of farmland and thereby the lending value. Today they are not so quick and ready with the numbers. The devastating farm crisis of the 1980's and the resultant general farm banking collapse have forced an inch-by-inch, case-by-case re-evaluation. That's probably the way it should have always been. Because beyond the obvious variables of proposed crop and/or livestock, prevalent weather, region, soil types, proximity to markets, and such, there are myriad other particulars which can have dramatic effect on the value of a given piece of farm property.

WHO ARE YOU

The most powerful OTHER variable is personal circumstance (and I'm not speaking of class or social position). For example; if you want to increase the size of your farm holding and the neighbor's twenty acres comes up for sale, it is safe to "suggest" that it will be worth more to you than someone looking for an investment. How do you factor in those two variables when deciding if the parcel is worth $400 an acre or $2,000 an acre? Operating farms in Lancaster County Pennsylvania have sold for twice and three times their justified farmland values to Amish families who are culturally, and personally, bound to try to remain

Your dreams will come to shape you and your world. If you dream and figure chickens all day long chickens will become your world.

within the communities to which they belong. And at the same time suburbia spreads like a pestilence putting altogether different pressures on those Pennsylvania farmland values.

On the flip side, there are hundreds of thousands of lovely farms with attractive buildings and good soils for sale at a fraction of their real value in areas suffering from large-scale out-migration. Portions of up-state New York, the upper peninsula of Michigan, Kansas and even the Ozarks fit that bill. In every case these cheaper farms are located in regions that are not close to large metropolitan areas. But these depressed farm regions do enjoy strong growing seasons, good soils, proximity to some markets and the well-established fabric of diverse farm communities. YET these are areas that, for the time being only, many folks don't want to live in.

That's the oh-too-simple truth of it. If lots of folks want to live in an area the land values go up. If they want to leave an area the land values go down. Lancaster County is a popular place for the Amish who've lived there for generations, for the small farmers who value the proximity to the Amish communities and other excellent market realities, and to the commuters who just want an acre in the country on a good road to the city.

But let's get to your situation. The question was something like *"how much should I pay for that farm I want?"*. We need to approach the same question differently for different folks because the suitable, and/or acceptable, price per acre will vary. So we need to figure out who you are. Let's oversimplify this and lump you into one of these categories:

A. Young adults, few assets, no tools (didn't know you needed them), no cash on hand, ineligible for conventional financing, limited or no farm experience, college education or part of one, no cultural or community ties to determine location (i.e. Amish, Native American, 3rd generation S. Carolina Tobacco), but an abundance of health, high moral fabric, enthusiasm, industry, creative intelli-

gence and good humor.

B. Middle aged adults, some assets (including tools), money saved, access to capital, college education, limited or no farm experience, used to convenience and comfort and high rate of pay, BUT absolutely MUST get out of the rat race and onto the farm, not as strong as you once were BUT know how to work, think you have a clear fix on what's important, long ago determination replaced enthusiasm, in search of the good humor that was lost somewhere in the city environment, think "creative intelligence" is a fancy way of saying "nut case".

C. Middle aged adults, very few assets (unless you count this year's vegetable garden and the cellar of canned goods plus the side of home-raised beef - Oh, and I almost forgot the old Chevy pickup is free and clear.), nearly a thousand saved, bad or no credit, no education worth mentioning excepting lots of practical hands on working experience including farming and ranching skills, lots of good tools (thought everyone knew they were important), used to working very hard for everything (except on Sundays), enjoy good health - humor -and outlook, value many friendships, already live in country on small rented place but always dreamed of a small farm of your own.

D. Nearing retirement or early retirement age, considerable assets (including equity in home and a stocks and bonds portfolio), $100,000 in nearly liquid form, pension and/or personal retirement income plan, no tools (or callouses), raised on a farm or ranch, adult life in city, concerned about health, education too long ago to matter, in desperate search for something long ago lost, suspect a return to a farm-like setting will bring back quality of life. Concerned about protecting finances as they represent old age security.

E. Middle aged or older. Used to be a commercial scale farmer but lost everything during crisis of the eighties. Slowly building back up. Assets include full range of tools and the complete knowledge of how to use them plus a dangerously clear fix on how not to get into the same financial mess again. Some education, strong family, good health, moderately good rate of pay working in agri-business industry. A little saved. Not much sense of humor, bitterness overrides, straight ahead intelligence wary of "creativity". Can't get the NEED to own your own farmland out of your waking dreams.

I hope you've picked up on the fact that there could be an infinite number of different categories. I'm sure that you don't fit any of these completely, maybe no one does. But this is an exercise in trying to show you how some seemingly insignificant things can have a huge effect on WHO you are and WHY you want a farm, and HOW much you might pay for it (not to mention how you might pay for it). The details of your personal circumstances and prejudices become very significant. A lot of these things bounce back to questions of WHY (or, ultimately, motivations) - which we TALKED about last chapter. And things like tools and fears will have an effect on your success.

It needs to be said that every one of the "example" people painted above COULD reach their goal to own their own farm. And I'll be borrowing from these examples in future discussions about HOW they can do just that. But here let's get back to the discussion of how much to pay:

DOLLARS PER ACRE

If your goal is to own a farm which pays for itself, and from which you derive your income and you kind-of fit in category A, C or E, you will most probably have to purchase a farm in an area where the land is cheaper. That doesn't limit you to just a few regions because almost every state of the union has within its boundaries areas of less population and low land prices.

The cheapest land I personally ever tried to buy was $85 per acre (offered for sale in 1987 and located in eastern Oregon). It had a small house, a falling down barn and corral, scattered pine trees, good native grazing land, a hundred acres of farm or hay ground. But there were two catches: you had to purchase the entire 1200 acres - and - it was twenty miles to the general store - post office and 75 miles to a city of any size. I resolved to try to buy it and called to make my "offer of terms" only to find out that this ranch property which had been on the market for six months sold the day before I phoned. It was purchased by a group of city doctors to function as a hunting retreat. They paid cash.

I know of remote farm and ranch property for sale for as low as $100 per acre, as I write this, but the new owner must be prepared to travel 60 to 75 miles to go shopping. (Please don't call or write me about this property because I'm sure it will no longer be available when you read this; which is an important point in and of itself - there is very little that is static about real estate availability and values.)

In the case of most all these cheaper properties, they are abandoned and often can be purchased with little or no down payment (more on that later). If the above figures can be used as a BOTTOM, up from that good farmland can be purchased for any price per acre from $100 to $10,000 per acre. The majority of farmland, sold in close proximity to markets and with a history of intensive farming, changes hands from between $500 to $2,500 per acre. Remember, true value, asking price and selling price may all be far apart from one another.

All farms are not equal when it comes to selling price.

If your goal is to enjoy a part-time farming experience and you do not need or wish to gain all of your income from the setup - and if proximity to hospitals and town convenience is important to you - a higher price per acre will have to be paid because you'll find yourself living where many others want to live. If you are in category A, C, or E this could pose a problem (but not insurmountable). If you are in category B or D this will not be a problem.

Acceptable rates per acre (if there be such a thing) can be, and often are, determined by figuring the production value of the land. If the land produces 60 bushels of wheat per acre (at $3.50 bushel) the annual gross potential receipt from that acre might be $210. Buying on contract and amortizing the purchase price of the land over twenty years (at today's interest rates) you'll end up paying off 10% of the purchase price (or mortgage principle) per acre per year. Add to that figure property taxes and calculate your cost of raising an acre of wheat (with or without a return to you on your investment and labor). It might look like this;

purchase price	$300 per acre
--	
yearly mortgage	$30 per acre
property taxes	$2 per acre
seed, tillage, fert., fuel etc.	$80 per acre
(equip, loan)	$50 per acre
sub total costs	$162 per acre
crop value/income	$210 per acre
return on investment	$48 per acre

To translate; a hundred acres of wheat with these numbers would result in $4800 annual potential return to you, 300 acres $14,400 annually. The numbers improve dramatically when you reduce the price per acre, raise the price per bushel, or reduce your inputs. Conversely, it is easy to see that there is no room to raise the price per acre or any other input without jeopardizing the financial balance.

Readers of the *Small Farmer's Journal* KNOW that we are always arguing the importance of diversification and appropriate technologies. The above scenario can be beneficially affected by reducing inputs with low cost appropriate technologies as well as increasing the return per acre of wheat grown by having livestock utilize the straw and help fertilize the acreage.

Imagine that you were considering fertile acreage suitable for market gardening: it is possible that this land could produce crops with a gross return of anywhere from $800 to $10,000 per acre. Of course the higher returns come from intensive perennial cropping with high offsetting inputs. It is a mistake to assume that blueberries or some other high return crop will automatically NET you more money. Net income is what you have left after all else is paid. And, by the economic measures we encourage you to use, what you have left must include waste products suitable for other purposes (i.e. fertilizer [manure] and stock feed [harvest waste]) as well as the measured gain you've enjoyed in increased soil tilth and fertility along with livestock gains. (Some will argue convincingly that the health, and enjoyment of the working farm family is also an asset which should be taken into the whole economic balance.) Crops that return less in a simple dollar measure may actually return more when you consider all these elements.

Market gardening acreage (with a history of this use) sells frequently for $850 to $2,500 or more per acre. And what makes a piece of ground suitable for market gardening has as much to do with proximity to primary secondary and tertiary population centers as it does fertility because fresh produce needs to get to the public quickly and easily. But, by the simple practice of drying or canning, shelf life can be extended and the farmer can justify being further from cities and thereby enjoy a cheaper land price (and, I must add, fewer suburban restraints on his farm practisces). To oversimplify: it is difficult or impossible to justify a purchase price of over $1200 per acre for any farmland if you depend on the farming to pay for the land *and* pay you back. Your success after the acquisition of the land will be affected by what you paid for it. You need to get the price down as low as is possible. If you pay too much for your farm you may be guaranteeing failure.

If you are new to this real estate game you have every right to ask how someone comes up with a given price for their farm - or even how the tax assessor comes up with a value for property taxes. The answer is that MOST prices and values are set after examination of recent similar property sales. In other words, if five people have recently paid an average of $1,000 an acre for similar farmland, in a set region, these sales establish a current defacto true market value. There is nothing to prevent you as a farm owner from asking ANY price you want when selling. But it is unlikely that you'll be able to sell for a price that is out of line with what other properties have sold for in the same area because most buyers shop around and are familiar with area real estate market pricing. There are, of course, exceptions to this rule but they won't include you most likely. If you are wanting to test the validity of a given farm price you should go to the county seat and ask the records department to let you see the "sales data files for

What is an acre of good hay ground worth to you?

real estate". Some counties may refer to these records by a different name, but they'll contain the same information. These records usually consist of real estate tax lot plot maps clearly identifying who owns which acreage. Either with those records or in a separate file, identified by the plot numbers, you will find when that property last sold and for how much. By checking these records you can find out when the seller purchased the farm and how much he or she paid for it. You can also check to see what neighboring farms sold for in the recent past.

That specific recent sales data is the important information. If the asking price is way out of line with those actual recent sales, you owe it to yourself to find out why or to give up on that parcel. There may be a perfectly legitimate reason why this farm is priced higher (i.e. gas or oil discovery or development, proximity to new zoning changes or developments, or other unusual extenuating circumstances) but there will be less reason for you to pay that much more than others have for comparable properties.

In some areas that have experienced increasing demand for real estate less than scrupulous realtors will approach landowners with no desire to sell with the suggestion that they might realize enormous profits from the sale of their property. In our own area it is quite common for a realtor to suggest, "let's list it and ask twice (or more) what you think its worth. If it sells we both win out big. If it doesn't you haven't lost anything." It can be confusing for you and seem somewhat dishonest but the fact remains that the asking price for a given piece of real estate may mean little or nothing. You may desire a certain listed farm property only to find out that the owner is far from serious about selling.

How do you learn the true status of that farm with the 'for sale' sign? Here are some important pieces of information for you to learn about a given selling farm. The answers will most probably tell you how serious the seller is.

1. How long has it been on the market?
2. How much is owed against the property?
3. Is the seller living on the farm?
4. Are there back taxes owed against property?
5. Are there any outstanding liens against the property?

Numbers 1,4,& 5 may suggest some urgency to the sale that would help you determine what is an acceptable *offering* price. It may also spell trouble in clearing up all the indebtedness with the sales transaction. We'll get into that in a future chapter.

Number 2 can be a very important piece of information when it comes to structuring a deal (more later) but for now it could translate to some inflexibility on price and terms.

Number 3 can give some clear indication of the up front cash needs of the seller. A family that needs to buy another home to move into will need more cash than an absentee landowner who is not looking to physically move. Absentee ownership might indicate that the farm property in question is but part of larger holdings suggesting that the owner may be best served by a land sales contract with a small down payment thereby avoiding or postponing some taxation. If you are able to give more of a down payment you will have a strong negotiating point to reduce the total price, especially if the seller is in great need. If you are willing to sign up to a creative land sales agreement with a slightly higher than standard interest rate you may be able to acquire land for little or no down payment, especially if the seller needs to unload the property but is not in need of a lump sum of cash.

We can never truly own a piece of land, we can only belong to it.

Keep in mind that the price you may pay per acre over time can be justified or offset in your budget by the amount you pay initially. If you are looking at the farm income to pay or help pay for the farmland, it becomes critical that your mortgage payments be structured to your advantage. We'll get into greater detail on this in a another chapter, but for now let me say that whether you have monthly payments, quarterly payments, or annual payments - or whether you split out interest payments monthly with an annual principle payment - or whether you set up up a mortgage with lower initial payments for a couple of years - or any other custom variety of mortgage - these determinations can have a tremendous effect on your financial solvency or insolvency. And there is NO one best way to structure your payment schedule, but there are several ways that a payment structure could hurt you. You need to have a clear picture of what you will try to raise and how you will market it and when you will market it to help you customize a schedule that at the very least will not set you up for hardship. For example; if your livestock and crop receipts will be coming in annually some time in October it would be doubly foolish for you to have a large mortgage payment in March. I say doubly because March will be a month of little or no receipts plus the possiblity of many expenses for seed, fertilizer, equipment, repairs, fuel and oil, etc, in preparation for Spring work. In such a scenario November would be a better month to have a big payment come due. (At the same time such a schedule provides yet another argument in favor of greater diversification and marketing creativity to spread the receipts - and expenses - out over the whole year.) All of this is to say that you may well be able to afford a slightly higher price per acre in exchange for some considerations in structuring the payment schedule. And don't forget the interest rate may well be a negotiable bargaining chip in the mortgage terms negotiations. Of course, if you can keep the price low AND get creative payment scheduling you will be better off yet.

Hopefully this discussion, and that of the previous chapter, will convince you that there is much to think about and prepare for in the process of buying a farm. Advanced preparation will be beneficial to you. And we've only touched on the beginnings.

When you buy another man's improvements give careful thought to their value to you and your plan for the design of your farming adventure. Make the beauty and utility fit your dream and your preference.

Here are a few off-the-cuff rules:

ALMOST ALWAYS TRUE

STANDARD RULE OF THUMB;
the fewer acres you purchase the higher the price per acre.
STANDARD RULE OF THUMB NUMBER TWO;
the closer you are to markets the higher the price per acre.
STANDARD RULE OF THUMB NUMBER THREE;
the closer you are to social services the higher the price per acre.
STANDARD RULE OF THUMB NUMBER FOUR;
the more other people want the property the HIGHER the price per acre.
STANDARD RULE OF THUMB NUMBER FIVE;
seven times out of ten the asking or listed price per acre can be negotiated down (or up - yes and perhaps to your advantage - more later).
STANDARD RULE OF THUMB NUMBER SIX;
realtors are NOT working for you or for the farm owner - they are working for themselves.

Please don't let all this material discourage you. The point in writing this is to try to help you succeed at your dream of buying a farm by providing information that will, hopefully, encourage you to be well prepared.

"I envy no man's prosperity, and with no other portion of happiness than that I may live to teach the same philosophy to my children; and give each of them a farm, show them how to cultivate it, and be like their father, good, substantial, independent American farmers - an appellation which will be the most fortunate one man of any class can possess, so long as our civil government continues to shed blessings on our husbandry."
- *Hector St. John Crevecoeut (a long time ago)*

"Agriculture is confessedly the most useful of all the Arts. Bodily health and activity of mind are eminently promoted by the exertions it requires. It is better calculated than other occupations for preserving the simplicity of manners, and purity of morals, which constitute the surest basis of a prosperous tranquility in states."
- *J.P. Bardley, 1801*

Chapter Three

*"Men stumble over the truth from time to time
but most pick themselves up and hurry off as if nothing had happened."*
-Winston Churchill

Measuring and Researching

In the last two chapters we discussed "Who wants to farm" and "The cost of farm acreage." In this chapter we will discuss finding and learning about your future farm property.

WHERE TO LOOK FOR FARMS FOR SALE

After you have decided upon an area or region the all too simple way to proceed is to read real estate advertising and talk with local real estate agencies. It is my personal experience that to limit your search to listed advertised properties is a mistake. Especially if you have made a firm decision on just where you want to settle. Often people new to this process will make the mistake of going to one real estate office, describing their needs and accepting whatever that agency offers, by way of information, as a complete view of just what is available. My wife and I knew we wanted to settle exactly

where we are and asked local real estate agencies for help in locating available properties. We were laughed at and sent away time and time again. We were told there was no such property available - period. We almost gave up when an out of town realtor returned our call and suggested we come by and pick up a map to a remote, barely accessible, ranch property EXACTLY in the area we wanted. He said it wasn't listed YET but thought it would come available soon. With his help we made an offer the very day it was listed and the rest is history.

Besides checking the ads and listings, drive the back roads, meet and talk with local folks, and visit the county seat and ask about acreage that might come up for auction sale because of unpaid back taxes. You are looking for people who MIGHT be selling soon, who are selling PRIVATELY without a real estate listing, who may be holding on to an abandoned farm property with the idea that no one would be interested in it, or who tried for a long time to sell and have tired or given up on the process. You need to leave your name and address with local banks and lending institutions with the information that you are looking to buy. You need to post little ads of your own stating what you are looking for. You need to snoop around.

You might find a farm or farms which are listed with real estate agencies.

You might find farms which are for sale by private individuals who prefer not to deal with real estate agents.

You might find abandoned properties which belong to absentee owners including banks or holding companies who would be willing to sell or trade.

Etc. Etc.

And each may require a different approach.

You may in your inquiries find realtors who have no listings that match your needs but who are quick to offer their services to you in finding a property. That is okay, HOWEVER if the realtor should ask of you that you sign a contract designating him or her as your agent I suggest it would NOT be in your best interests to do so.

Once you've found a farm you're interested in, the work really starts. Now it's time to research it and measure it against your needs and circumstance. Buried in this chapter is a master checklist which will hopefully ease what can be a trying process

After locating a farm property

WITH A REALTOR

If the farm you've found is listed with a real estate agency you might be able to have them assist you in your work. But remember: REALTORS, and Real Estate Agencies, WORK FOR THEMSELVES, not for the buyer, not for the seller. They will almost always push to close a sale as fast as possible whether it's to you or someone else who happens along. If you end up buying the wrong farm, or the right farm the wrong way, and find yourself in difficulty the realtor will still get his or her commission. Your best interests are not necessarily theirs. The honest, easy-going, intelligent real estate agent is rare. Be on your guard.

So first, let's proceed as though a realtor were involved:

1.) Get a copy of the listing agreement.

2.) Ask if a recent title search has been done on the property and, if so, get a copy of the results.

3.) Name and address of owner if somehow missing from listing agreement.

WARNING: DO NOT LET A REALTOR TRICK YOU INTO A PREMATURE 'EARNEST MONEY AGREEMENT' BY SUGGESTING IT'S THE ONLY WAY THEY CAN RELEASE THE ABOVE LISTED DOCUMENTS. SIGN NOTHING WITHOUT REVIEW BY A KNOWLEDGEABLE THIRD PARTY (preferably a lawyer).

If a realtor is unwilling to provide the above information, suspect a problem with either property OR realtor OR both. (A good realtor will understand the necessity for this information and will work agressively to anticipate all the questions you might have so as to be prepared. This is to the realtor's best interest because it expedites the inquiry process and can speed a sale without undue hassle. If you get lucky you'll find such a realtor and cut your "investigation" time way down.)

NEXT

Assuming you've got the documents, look for the answers to these questions:

a.) Who actually owns the listed property? What is the asking price? Does the seller require cash in full? Will seller allow a percentage down payment with owner financing? What down payment amount is required? What nature of owner financing (i.e. land sale contract, deed of trust, etc.) is acceptable and sought? Is there an existing mortgage against the property and is it assumable or transferable?

b.) What is being bought? Exact acreage? Some paper evidence of any conditions as in guarantees of production levels or statements as to 'sold as is" (more on this later). Improvements specified? A complete listing of all items to be included with purchase as well as those not going with property (i.e. portable outbuildings, irrigation equipment, above ground storage tanks, etc.)

c.) Is anything currently owed against property? What manner of contract and security is involved? Are there any liens currently held against the property and by whom? Any collateral pledges as guarantor for other transactions? (Later, we'll talk about STATUS of above.)

d.) Any easements recorded (more later)?

e.) Any rights withheld (i.e. mineral or timber rights not transferring or transferrable and easements related to same)?

f.) Water rights recorded and filed? Age of same (more later).

g.) Property taxes? Any in arrears?

h.) Relevant districts? School, etc.

i.) Any conditions for viewing property and improvements? (i.e. lengthy advance notices and restrictions on subsequent revisits during purchase negotiations.)

In a few paragraphs you will learn the significance of the above questions. (Just because you have some paper evidence of this information you must be careful not to BANK on the veracity (or accuracy) of it. This is why you will insist on **escrow closing** with a fresh **title search** and **TITLE INSURANCE** when, and if you close the transaction [complete the purchase of the farm.])

After reviewing this information return to the realtor and request these additional facts:

4.) Pertinent local zoning restrictions, (i.e. dust, noise, livestock, restrictions). *Now is the time, NOT LATER, to discover that your dream of a broiler or dairy operation would not be allowed by local ordinance due to dust, noise, manure and fly concerns.*

5.) Neighborhood information: Churches, Schools, distance to markets, fraternal organizations, proximity to fire protection, etc. *Now is the time, NOT LATER, to learn whether or not the general community might be at odds with your religion, value structure or appearance.*

6.) Cost of utilities, *(especially if electrical service is required for some important aspect of the existing or proposed farming operation such as irrigation, refrigeration, etc.)*

7.) Current fire insurance coverage. *(If the conditions of the buildings or the absense of a fire district rule out fire insurance coverage you should know this in advance especially if third party financing is required. Often a lender will require fire coverage to protect the security interest, if none is available it MAY result in financing falling through. It is always best to understand this limitation at the beginning of application processes as it may save a lot of headache and expense down the road. We have no fire insurance coverage on our ranch because we have no fire district, however, we were able to finance the purchase with a lender because the borrowed amount was far below the full value of the property.)*

8.) Wells tested for water purity. *(Contaminated water source(s) can make a particular farm seem far less attractive and even hazardous to your family. It's a good idea to get this information early on. The tests are usually inexpensive.)*

Many realtors will baulk at above requests often for no more better reason than because they are lazy. If the information is provided you casually (as in *"oh, sure yeah we checked all that and its fine..."*) **doubt it** and require the realtor to verify it in writing.

That done, you will have to dig up a little more. Research the answers to these added questions yourself by talking with the seller, neighbors, experts, and local business people (i.e. farm supply companies and feed store owners).

9.) What are the resident soil types? Most recent soil type analysis? When was the last soil fertility test taken and by whom? *(This information will be valuable to you in determining what crops may be initially suitable and the degree of soil rebuilding you will have to plan for.)*

10.) History of serious livestock contagions? *(Make a point of checking with the local veterinarians to find out if there have been any serious contagious livestock disease outbreaks in, on and around the farm in question. It could be disastrous for you to move your valuable breeding stock to the new farm and have them all die of an avoidable or preventable disease.)*

11.) History of farming practices? Nature and extent of agri-chemical applications over last three years? *(You will benefit from learning whether or not this farm has been organically, biodynamically or chemically farmed and to what relative degree. Even if it is your choice to farm chemically it is still important to know if large amounts of certain substances have been recently applied and what that might mean to your future plans. And if your plan is to farm organically, chemical residues will have a negative effect on those plans whereas a history of poison-free practices will be worth money to you.)*

12.) Structural condition of buildings? *(Will you be faced with some immediate and expensive repairs?)*

13.) History of orchard, woodlots, ponds, physical hazards (i.e. dumps, abandoned wells, etc.). *(Any and all information about the farm or ranch in question will be useful to you. And learned early may have an affect on your decision to purchase and acceptable purchase price.)*

The information you gather with this list will explain itself for the most part. Early in the process you may find out things that rule this or that farm out of consideration.

Often even good realtors (and there are many) will need to make a determination as to whether or not you are a serious qualified buyer. Why waste time with you otherwise? There are a lot of lookers and dreamers out there with no intent of ever purchasing. Having the money doesn't automatically put you in the category of a qualified buyer. Coming with a clear idea of what you want and a shopping list of intelligent questions will put you there. If you are a "qualified" buyer the good realtor will want to provide you with all the information you require no matter the outcome. Because the "good" realtor will recognize in you someone who's prepared to buy that "right" place. Someone they can sell to eventually, especially if they earn your trust.

WITHOUT A REALTOR

Now, if a realtor is not involved and the farm is being sold by the owner you still need most of the same information and it may be more difficult for you to get it. Above, you've gone through a list of things to ask for. Below, in the same order, is the identical list but worded, rather, as what you need to know. Use this as your master check list.

1, 2, & 3.) Is the property listed, contractually, with any real estate agency? *(Sometimes the owner is given, or takes without lisense, the opportunity to sell his property himself even though a listing with a realtor exists. It is important to know this as it could certainly gum up any purchase process.)*

Does any person, company or entity (i.e. estate or attorney) control or stand to gain from the sale? If it is listed you will find on the listing agreement answers to most of the questions under A.). If it is not listed you will need to dig up these answers.

    A.) Who owns it? Asking price? How many acres?

Of course you need to know the name and address of the person (or company) selling the property. It doesn't hurt to know something about them. For instance, does the owner live on the farm and receive any income from it (valuable to know)? Or does the owner rent it out, and for how much? Does the farm belong to a bank or insurance company? Did they get it through foreclosure? Are they interested in carrying a contract for the purchase? The personal or business circumstance can have an enormous effect on the purchase of the farm (i.e. does it belong to a bank that is burdened by foreclosures and MUST sell?).

How must it be sold, all cash, % down, assumed balance?

    B.) What is being sold?

What goes with the sale? What does not? Do not assume the irrigation pipe and portable hen houses are selling. Find out if they go with the deal, and in writing. Find out what DOES NOT go with the sale. Find out if any rights or easements are being reserved. Don't wait until after you own the farm to find out some company has the right to pull their drilling rig into your field and start pumping gas or oil.

    C.) What is currently owed against the land? Is it assumable?

"Assumable balance" *might* mean that the seller still owes money on the land and you can step in and take over the payments if you wish, and if you are eligible. This can be most advantageous in structuring a purchase. It can also be deadly. Understand this because it is critical! For instance, if there is a FmHA loan against the property will YOU be eligible to assume its balance, will they (FmHA) allow it? Before putting any money on the line understand these things and KNOW them to be true. UNDERSTAND ALL CIRCUMSTANCES and get legal council before signing to assume someone elses debt! When looking at what is owed against property, it's important to discreetly determine the status of that debt. If the seller is near foreclosure and faced with imminent possibility of added liens being placed against property (i.e. back taxes, etc.) best intentions may be foiled before a transaction closes. And worse you may find yourself saddled with a nightmarish process of lien and foreclosure started some time ago and which only actually commences immediately after you THINK you are the proud new owner. An attorney and a title company can be valuable in helping to untangle such a situation to provide a clean transfer.

On the same subject: be wary and well informed of any secondary operational agreements you will be expected to conform to if you do assume existing financing especially with some sort of federal or quasi-federal farm credit program or agri-business production credit company. You may be signing a contract with fine print that states you must grow and sell only corn and only to BIGSHOT CORN INC. or you might be setting yourself up as a "cooperator" with a government program that stipulates you must do what the friendly bureaucrat says or else.

D.) Are there questionable easements?

An easement is a titled right to some person or entity to cross or access your property usually for a specified purpose. For example, it's common for a power company to have an easement to work on the power line. Those are standard and not a problem. Question easements across your property to access landlocked holdings. They might legally become deeded roads in some cases. If an easement seems suspicious don't accept it until it's been clearly understood.

E.) What rights, if any, are withheld? Are there any stated Deed restrictions?

Ideally you want to purchase the land with all rights intact. Sometimes a seller does not hold mineral or subsurface rights to his property - or wants to keep them. What this can mean is that someone else may be pumping natural gas out in the middle of your lambing pens!

This can be such a complex concern with land purchase that the specific variables have to be reviewed with knowledgeable council. Talk with an attorney and BE CAREFUL!

F.) Are there any registered, recorded, proven water rights?

Recently this has become the subject of nightmarish concerns in arid states where irrigation is critical. It is possible to have a recorded water right which has lapsed for lack of use in which case it is sometimes stated that the right is no longer "proven". It is possible to have a proven water right but be denied by the state access to the water because drought or environmental concerns limit the amount of water to be taken and your water right is "younger" than neighboring farms. Older rights take precedent. There is a lot to learn about this critical issue. It is enough to say you should research the subject for your area and make sure you understand the situation where you choose to buy land.

G.) What are the property taxes?

Are they behind? What is the likelihood of a substantial increase?

H.) What districts does the land reside in?

In some areas there are jurisdictional districts for irrigation, land use, schools, etc., that can have very specific affect on the landholder. Find out about these.

I.) Will the seller, or realtor, allow you access to view the property without undue restriction?

4.) Are there local building and use restriction zones (county, city and state) which will affect your plans for this property?

Currently many states are drafting "right to farm laws" which may build into the regulations designated buffer ag lands around high-density suburban areas with restrictions against dairying, nighttime field work, manure storage, chemical applications, etc., etc. Find out if it applies to you. Also building restrictions may surprise you when you go to put up a second house for your family or help.

5.) How far is it to church, schools, markets, etc.?

How strong a community is it? Do the farmers appear to share your philosophies and values? Are there lots of attractive small farms? Are there only one or two neighbors? Personal considerations figure very high here.

6.) How much do the utilities cost.

Any legal restrictions to producing your own power?

7.) What is the current fire insurance coverage on the farm buildings?

This can be useful information in determining conventionally measured building replacement values.

8.) Is the water good?

This is important. Get samples of thedrinking and irrigation and pond waters tested. This can save a lot of grief and hardship later.

9.) Have the soils been mapped?

Check with appropriate county offices to find if the soils on this farm are classified and mapped. This will help you determine suitability for certain crops. Also find out if there's been any recent soil tests and the results. You may want to have some done to help in your determination.

10.) Has there been any serious, contagious livestock diseases on the place in 10 years?

Check with local veterinarians and extension agents. There may be an important reason those poultry houses are empty.

11.) What has been the farming history of this place?

Find out what has been expected of the land. Has corn or tobacco been grown forever? Any conservation practices? What chemicals, if any, have been used? What has been the rotation? Livestock history?

12.) What is the structural condition of buildings?

Examine foundations, roofs, wiring and plumbing for BIG future surprises. Know NOW what the problems might be.

13.) What is the history and specific identities of the orchard, woodlot and pond?

Also the same of any physical hazards like dumps and abandoned wells. It will be helpful to know age and varieties of fruit trees. Also helpful to know any maintenance history on the ponds (if any). It may be critically important to know what was in those barrels now stacked in the farm dump and where the abandoned wells are.

Well, that's the master list. It may not include everything, but I guarantee that answers to these questions will send you a long ways towards a decision about whether that farm or ranch is right for you.

And remember, it may be right for you but just not something you can buy.

There are some remaining considerations but they can be so various that a list is impossible. They include your likes and dislikes. But more important they include your actual system design and expectations. You have to decide if this farmland will work for your mix of crops and livestock. You have to decide if it will work in your master plan.

Next chapter we'll cover Financing, Bankers, Realtors, Offers, Purchase Agreements and Contracts.

Again, please don't let this morass of information discourage you. This is meant to help you avoid pitfalls and succeed with your dream of buying a farm.

Jack Carver and team harrowing the market garden in New Hampshire.

Copy the checklist below and take it with you when you meet with a Realtor or seller or both. Ask these questions and, even if answers are promised in writing later, listen carefully to the verbal answers.

MASTER CHECK LIST

Is the property listed for sale with an agency?

Who owns it?

Asking price and terms?

What exactly is being sold?

What is currently owed against the land? Is it assumable?

Are there questionable easments?

What property rights are being withheld?

Are there any water rights?

What are the property taxes?

What districts does the land reside in?

Any restrictions to viewing the property?

Are there local building and land use zones?

How far is it to community services?

How much do the utilities cost?

What is the current fire insurance coverage?

Is the water good?

Have the soils been mapped?

Has there been any serious, contagious livestock disease on the place in 10 years?

What is the farming history of the land?

What is the structural condition of the buildings?

What is the history and specific identities of the orchard, woodlot and ponds?

    And hazards?

Chapter Four

"If a man would enter upon country life in earnest and test thoroughly its aptitudes and royalties, he must not toy with it at a town distance; he must brush the dew away with his own feet. He must bring the front of his head to the business, and not the back of it."

- Donald G. Mitchell, 1863

Money Changing Hands

As we suggested at the opening of this book: Some of you will find this writing simplistic, but evidence is that we need to attack these questions at the most basic level, not just because we have new people looking for introductions but also because many of us can benefit from having our motivations and knowledge questioned and reevaluated.

The price of a farm starts a process that can lead to a greater or lesser total cost. You have a lot of control over the outcome. But you need information. Hopefully you will find some here.

I strongly advise anyone new to these pages to read the first three chapters of this book. Also, as wide and varied a subject as this is, it is impossible to include all elements. Although we're going over specifics, it is the tone of doubt, inquiry and insistence that is most important to communicate.

Because of **Small Farmer's Journal** we hear from a number of folks buying their first farm. Far too often we get sad stories of tragic messups in the final stages of money changing hands. Can we learn from our mistakes? Yes.

Last chapter we covered some critical preliminary ground covering the researching of properties and listing agreements.

Assuming that you've found a farm you want to buy, next you'll need to determine if you can buy it. If you have sold your property, and/or saved your money, and have the means to buy the farm you are sitting pretty. Best of all possible worlds - congratulations. However, don't stop reading. There are plenty of pitfalls that can surprise even you. (Perhaps especially you, since unscrupulous people may identify you as a more than likely culprit.) Piece of advice: How much money you have, in what form, and where, is nobody's business but your own.

If you do not have the full price of a considered farm, in cash or any other form, you will likely have to look to financing (or borrowing - and they are not necessarily the same).

But before we get into that, there is an important - inbetwixt - option, trade. If you have property that you want to sell, but haven't yet, consider trading it for, or towards, the farm. It isn't necessary that the farm owner want your property. A realtor with some intelligence and hustle will have the information and means to possibly locate a third property (one the farmer wants) that can tie the deal together as a three-way trade. Oft times the values are different and money also changes hands. But it can work out quite well. It's too complicated to go into great depth and detail but trading is an important option worth looking into.

FINANCING and/or BORROWING

If you need financial assistance in order to buy the farm, here are some basic rules which vary some from region to region, and with differing particulars.

(1) Real Estate Mortgages are usually long term. Long term real estate mortgages are usually written for 10, 15, 20, 30, and 40 year contracts with 20 years being most common. Often contracts can be written with a payment schedule amortized (calculated) for 20 years or so but with the contract maturing at 10 or 15 years. What this means is that you will have a balloon (or considerably larger) payment to pay at the end of the contract. Some big banks prefer this program because it gives them the opportunity to gain interest points and fees if, and when, you choose to rewrite the mortgage to extend its term. For you it's a device to keep payments lower for the first years. An important procedural tool.

(2) Interest rates on long term mortgages can be fixed or variable. Fixed means that you pay the same rate for the duration of the contract whether it's 9% or 20%. Variable means that the rate goes up or down by a formula usually tied to fluctuations in the prime rate. If you have an opportunity to get a mortgage at 10% or less it may be wise to go fixed. If the prevalent rate is 18% at the time of your purchase it may be wise to go variable. You'll have to do some math to get the answer that's right for you.

Interest rates can be a strong bargaining point in contract negotiations. If you're willing to offer 11% rather than ten, the seller, or the banks, may be willing to extend the term for more years, reduce the price, or throw in some equipment. The opposite might also be true. If you insist on a lower interest rate you may have to accept some other change that might raise the price or cause you some inconvenience.

(3) I am of the opinion that a well crafted purchase contract directly with the seller is far superior to dealing with any bank or lending institution. A careful

review of banking news, relative to farmlands, over the last 15 years will show a trail of stupidity, greed, avarice, fraud and vulgarity on their part. There are exceptions but they're becoming fewer and farther apart. With, what we out west call, a "land sales contract" (or seller financing) both the buyer and seller can be rewarded.

(4) You, the buyer, must either have your own attorney draw up the contracts or at the very least have your own attorney review the contracts BEFORE SIGNING or MONEY TRANSFER!! Do not use the seller's attorney, the bank's attorney, or the realtor's attorney! You want professional help in your employ and on your side.

(5) Understand the important differences between an "earnest money agreement and offer" and a "contract to purchase" (more later).

(6) Federally guaranteed loan programs for farmers (i.e. FmHA) were originally conceived to assist beginning farmers who could not get funding elsewhere. They became, and evidence suggests they still are, a money tree for wealthy cronies of elected officials. The extent to which these programs have hurt farmers is evidenced by the long hard hours of legal advocacy work they have generated. Our advice is to stay away from these programs. (We're not able to speak from experience about G.I. loans. You'll have to research those.)

(7) Insurance companies and fly-by-night mortgage companies make loans to purchase farmland. Their rates are not competitive and their motives are suspect.

(8) Lenders usually loan 50% to 75% of the land value.

To summarize: Go for a "land sale contract" with seller. If not possible, be very careful dealing with the other options. Go for a 20 or 30 year amortization with a 10 or 15 year balloon at the lowest possible long term interest rate of the day. Have your attorney involved.

REALTORS

You've, hopefully, read the previous comments about realtors. It is enough to say here that they are human-beings in a pressure situation selling land for a commission. Unscrupulous or not, they get paid a percentage IF the property

It is up to you to buy your farm safe and clean. Don't let mistakes in the purchasing process threaten your tenure on the farm and detract from those memorable moments like your child's first horse back ride.

sells. They are NOT working for you, the buyer. They are working for themselves. Remember that at all times.

When you find the property you want, you have to get aggressive and take the initiative away from the realtor. Take charge! Read last chapter's piece on preliminary research. Know what questions to ask.

Then, when you are ready to make an offer, understand that this is the most important stage in the negotiations to buy. It will be difficult or impossible for you to make any significant changes in what you offer (price, terms, interest) so make sure you have carefully thought through your stated preferences. You should be prepared to put up some money, to be held in escrow, as a show of your good faith. This is called earnest money. Commonly anything from $500 to $5,000 is offered. This money accompanies an "earnest money agreement and offer to purchase." In this document which is usually drawn up, on forms, by the realtor (but which can be drawn up by your attorney or can sometimes be purchased as a blank form at larger stationery stores) you will state your purchase offer including:

1. How much you are willing to pay.
2. How much down payment you offer.

3. How it's to be financed / what % rate / length of contract, etc.

4. What you want included with purchase.

5. That this offer is subject to the contingency that a title search will show clear title, that review of applicable local regulations show no restricted use, and that financing be forthcoming (if other than seller contract).

6. That this offer is subject to sale and purchase of said real estate being accepted, completed, and closed by a certain date.

7. And any other specifics that are germain.

When this is drawn up make sure that the realtor gives you a very specific receipt for the earnest money check you give over. But first it is important to have your attorney review the earnest money agreement before you sign and transfer any funds.

I have, in the past, had realtors try to tell me how to word the offer and what price the seller might take. Remember who the realtor works for. The realtor gets more money if the price is higher. You INSIST on your amounts and that the offer be made. The realtor can get into real legal trouble if he or she refuses to submit your offer.

And perhaps this is a good place to insert yet another insistence that you get an attorney to help you. An attorney familiar with real estate contract law can breeze through some of this paperwork in no time at all and $150 spent might save you your entire life's savings. Remember if you hire an attorney to represent you, that attorney represents you.

After the offer is submitted the seller, of course, has the right to refuse or make a counter offer. Once you've reached agreement on price and terms I strongly suggest your realtor be involved in the drafting and at least the review of the purchase contract.

Remember that this document, the Purchase Agreement, will be your primary tool of enforcement for any breaches, non-compliance, or enclosures.

If the purchase of the property includes appliances, or farm equipment, or crops, be sure that all is spelled out in appropriate addendums to the contract. If the seller is supposed to have something done prior to your occupancy be sure it's spelled out.

I once made an earnest money offer on a ranch property that was accepted. As we waited for the drafting of the contract, I went out to the property for a picnic to find a big crew cutting down every tree and loading log trucks in a hurry. They were told by the seller that they had 10 days to log it off before the contracts were signed. The earnest money agreement protected me, and we backed out of the purchase - disillusioned but without monetary loss.

Next make sure that the escrow company to handle title search, insurance, closing and possibly contract collection is legitimate. Your attorney may be able to help you here.

While waiting for the contract get a copy of the preliminary title report and go over it with your attorney. It may save some surprises later.

Escrow is a special, legal, "third party" service provided so that critical negotiations, contracts and transfers between two or more parties can be transacted without problems.

Closing is a term, in the case of real estate, used to describe that time frame and process during which paperwork, monies and transactions are all enacted.

I recommend that CLOSING always be done in ESCROW!

Title insurance is what it says. A company doing business providing insurance on titles will research and provide the ownership status report on a piece of real estate and then be willing to insure that what they report is accurate.

You must get TITLE INSURANCE.

A man I know gave someone cash for 40 acres and was given a written deed. No contracts, no closing, no escrow, no title insurance or reports. Soon after someone came to him claiming he owned the land. After months of investigation it was found that land had been sold three times by the same seller and actually belonged to a bank which held security against it for a loan. All three people had to individually sue to get their money back. None ever did.

And don't be surprised to find out that all these steps and precautions cost money. Normally the buyer and seller share, 50/50, the closing costs which include contract drafting cost, title search and insurance, plus escrow and collection. If there were any back taxes owed against the farm they are usually cleared up within closing. Be careful that someone doesn't try to stick those, or half of them, on you. Remember, it's the sellers obligation to provide you with unencumbered property.

Well, we've covered a lot of ground in a hurry. There is just no way we can talk about every possible detail. What we meant to do with this is give you the kind of background that will improve your chances for a successful purchase.

The purchase of your farm, a difficult and sometimes confusing process, can and should be completed in a clear, straight forward, and honest manner. Don't build regret into every future day's farm labor by allowing yourself to purchase a farm in a questionable process or procedure. Buy it right, buy it clean and then move on to making it work for you.

You can't imagine the joy, tinged with terror, that comes when you have finally bought that farm. "What have I done?" and "I can't believe it's really true" are sentiments most often felt. If you do your homework and exercise intelligence and patience you will be able to get full enjoyment out of this life shaping moment. As you watch your neighbor's buggy go past your new farm, pinch yourself to know that it's not a dream.

Chapter Five

"And who can gradually claim the right to point to all accumulations of small gestures over the days and months and years that bloom into something as quietly satisfying as a field of garlic or a mud house or a small farm, and all that which has been labored for, not simply bought or found or taken."
        - Stanley Crawford*

AFTER THE FARM IS BOUGHT

In the first four chapters we've presumed to take you through the temporal steps of buying a farm. And we began this discussion with the process of deciding what farm you wanted. In the last segment we finished the actual purchase scenario. In this final part of the series I'd like to touch on some critical considerations which just might help, over time, to determine your purchase as a success.

If, after a while, you come to judge your purchase a failure it could be because you didn't take the requisite precautions and got snookered. Or it could be for altogether extraneous, or outside reasons. Either way we aren't going to concern ourselves with that now. But, it is possible that the purchase became a failure because you either;

1) could not make an adequate income from the farm to justify (or "pay for") the purchase of it or repay operating loans.

---

*HOEING GARLIC 2/92 Parabola Magazine, from A GARLIC TESTAMENT: Seasons on a Small New Mexico Farm, copyright S. Crawford 1992. Edward Burlingame Books/Harper Collins, New York.

Or -

2) you dislike the nature of the work you found yourself doing. It isn't what you thought it would be.

Or -

3) you like the work but you can't handle it all.

Or -

4) most important: you couldn't afford the livestock, equipment, and/or seed etc. that you deemed necessary to give the venture a real try.

All these possible problems can be addressed right after purchase, during your first days as farm owner. But in truth, they should have been factored into your considerations from the beginning.

Will the milk this cow produces sell for a high enough price to pay your farm bills?

For example: Intelligent inquiry and computations should have been made, from the outset, to determine if beans at 18 cents a lb and milk at $10 cwt would add up to revenue adequate to handle debt service, taxes, operating expenses and a living wage.**

And that same inquiry should have gone far enough to suggest to you that if you lock yourself into beans at 18 cents per lb, and milk at $10 per hundredweight you've made a big mistake because you've limited your options from the very beginning. Your farm has to be special, unique, and alive in ways that

---

**(Most farm economists will hasten to save you the time and tell you it can't be done - but they're academic ostriches who see only in terms of common denominators when considering highest costs and lowest income. And extension agents are hide-bound to "enterprise data" created by those store-bought ag economists to justify the rural terrorism of our federal U.S. and Canadian farm programs. How can we accept advice or counsel from the government when it continues to work to destroy the farm community? We must trust our suspicions and instincts and go to successful individual examples and small farm advocates for counsel and direction. We must come to accept that success can be affected more by a romantic outlook than by abstract accounting or modern measures of efficiency. What you do is important, how you do it is also important but WHY you do what you do is most important of all.)

industrialized agribusiness does not allow.

If you're saying "okay, tell us those ways..." Good, you're listening. But I can't (or won't) tell you those ways here and now because that would be a sidetrack. Just, please, hear this: The ways are out there, they are as varied as the people using them and they are as various as the blades of grass from North Dakota to Texas. And those ways might give you whatever level of cash income you need to pay the freight but you have to meet the train at the station, so to speak. You have to take a hard look at what is important in your life and practice a true frugality and thrift. That doesn't mean doing without. It means appreciating what you have and understanding how what you value comes to shape your life.

The excesses of this last half of the twentieth century have made such consummate gluttons, and lazy bums, out of many of us. We fill our lives with such a lot of gadgets, and trash and services that we do not need. I am reminded of a couple that moved to an out-of-the way Iowa farm and were upset at not being able to find a garbage service to pick up their trash every week as had been the case in the city. They had difficulty making the farm pay for plenty of reasons but close inspection of their books disclosed that 65% of their personal living expenses were non-essential and wasteful (i.e. timeshare payments on a lakeside condo, membership in a video-of-the-month club, jewelry purchases, mobile phone service for the new pickup truck, farm consultation services, payments on a radar dish for television, payments to the neighbor's teenager to wax cars and mow the lawn, etc, etc.) Just makes me wonder why they moved to the farm in the first place. Fact is that folks these days cannot pay the bills on this kind of silly frivolous greedy lifestyle even with high paying city jobs.

But we're getting off to the side of what we want to talk about.

FIRST THINGS FIRST

It is common for people to want to make some immediate changes in their new farm. Don't. Give yourself and the farm a little time - or at least a closer look.

You might think that your new farm is fenced all wrong, or that a certain tree is in the wrong place, or that a wet area would be better drained, or that this

"Look at what is important in your life and practice a true frugality and thrift. That doesn't mean doing without. It means appreciating what you have and understanding how what you value comes to shape your life."

gully would make a good pond site, or that a depression in the road should be filled, or that the old sheds should all come down right away. Well maybe you're right on all counts. But, just maybe, you're wrong. Your casual momentary observations cannot take into consideration how all these aspects of your new farm fit into the year and operation of that farm or into your own future needs. Until you have to move the cows or sheep from field A to field D during the summer cropping season how can you be expected to understand the true value of what you thought was a wasted piece of field for that lane? And maybe when you drain that wet area you're going to have an effect on plants all around it which shelter and feed small animals that figure into the control and balance of rodents that might grow in numbers, without predators, to cause expensive damage to the hay and crops. How were you supposed to know that damming that gully would be a colossal mistake when snow melt created a massive runoff that filled the pond and then cut the dam out and flooded downhill improvements? And after taking down all those old sheds you've found yourself short of time and money to replace them with better facilities, and now wished you had just fixed them up for the time being.

This is not an argument against action. It is just an admonition to take it easy in the beginning because you don't really know that farm, yet. And many of the things you've inherited in its design have been the result of someone else's trial and error. They may not be perfect but they just might be there for a good reason. Many is the new farmer who saddled himself with big avoidable cash outlays because he was in a hurry to "improve" his place and made some costly mistakes.

The first thing you should do when you've bought a new farm is take a vacation! Right there, on that farm, vacation. Relish the place, visit with it. Walk every corner of it. See it from every angle. Court it like you would someone whose love you want returned. Relax in her embrace. Sit on the fence and tell this new place all your dreams. Sit under a tree and listen to all this farm can tell you at a glance. And realize how truly insignificant you are in the life of this place. That what you've really bought is a chunk of time with this place. Think about how awesome your responsibility is to the shape and color of this place for the foreseeable future. This is no time for arrogance. This is the time for relaxing in humble reflection about the future of this place and your time with it.

Another frequent mistake born of haste is your choice of counsel or advice. We've touched on that somewhat, already. But allow me to add a little.

You are justifiably anxious about your new farm purchase. You want it to work. So you go to those places and people who have their shingles out to sell you products and services and advice. 95% of the time these people will give you the "commercial" line of what farming is all about. And their advice will have you working towards an agri-business approach or deciding that you made a mistake in your purchase of a farm. Both conclusions are unfortunate and unnecessary. 4% of the time you will have smiling fast-talking folks selling you schemes for making a fortune off your new farm in just a matter of months but it does require an initial investment...New farmer beware. A simple rule of thumb is that if it sounds too good to be true it probably is. And if you're lucky, occasionally you will find yourself talking with happy successful small farmers who are too busy to offer you any counsel, too humble to figure they have anything to offer you, and maybe just a little protective of what they have going on. This is the source you should cultivate. But keep yourself open to go your own way.

"The first thing you should do when you've bought a new farm is take a vacation! Right there, on that farm, vacation. Relish the place, visit with it. Walk every corner of it. See it from every angle. Court it like you would someone whose love you want returned. Relax in her embrace. Sit on the fence and tell this new place all your dreams. Sit under a tree and listen to all this farm can tell you at a glance. And realize how truly insignificant you are in the life of this place."

That's what the "independent" part means.

If you found yourself flirting with insolvency and unhappy about the nature of the work you're doing, you'll find yourself loosing sight of WHY you chose farming. That's because the HOW of it has taken over. If you came to farming still worshipping the goddess of convenience you lose. Farming is about craft and working with your hands and feet and whole body to orchestrate a process of creation, cultivation and harvest. Whenever you allow the process to become industrialized you lose some part of your connection to the process. You lose part of why you do it at all. Whenever you court convenience for convenience's sake you strip some part of craft from the craftsmanship. We farm not because we want to avoid labor, we farm to save labor for ourselves, save

"If you've found you don't like the work give a close look to how you are doing the work. Perhaps you have an opportunity to change the system and make it work for you."

labor for the strength and self-determination it gives us when we are in it.

If you've found you don't like the work, give a close look to how you are doing the work. Perhaps you have an opportunity to change the system and make it work for you. I'm remembering the late Parker Sanborn, of Maine, and how he quit making hay on his Jersey dairy farm, stopped feeding expensive supplements, cultivated the best of pastures, selected cows for longevity and milk production on pasture - and fell completely in love with the day to day nature of his work. And for Parker the economies, and the peace, followed suit.

Although it may not be for everybody I have to add here that for me all the past ugliness of farming seemed to slip away when I switched to working draft horses. When I worked tractors I hated every minute of it. I just wasn't suited to the mechanics, the fumes, the vibrations, the noise. After twenty some odd years I still relish every moment with the horses whether plowing or mowing or whatever the job. Someone once observed of me that it looked like I was a farmer to have an excuse to work the horses. That's going too far because I love every aspect of farming but it certainly pays me big dividends to be so rapturously happy with HOW I farm.

So here you are with a brand new farm, in a position to make decisions about how you farm. Choose carefully my friend, I hope you err on the side of what you are attracted to, rather than on the side of another "practicality". That is what I mean when I say that romance is important.

Another very big mistake folks make when they buy a farm is to set the stop watch to go from the first day. When you first decide to be a farmer you must somehow learn to realize that the process of building a farm is as the quote at the beginning of this piece has said, the result of "...accumulations of small gestures...". If you are determined to have all the fields just so, the livestock right now, the implements and buildings done on day one, and cash available to pay for each day's passing - you are courting certain disaster no matter how rich you might be. For, if you haven't the money to begin with, borrowing it will put almost insurmountable pressure against you and your future efforts and deny you the true wealth that comes of building the "accumulation" of small gestures slowly. And if you are rich, throwing your money at the process will remove you even further from the earned sense of place and accomplishment.

It is here that you must bring real intelligence and innovation to bear on the job of getting the farm started and keeping it solvent. Notice I said "keeping" instead of "getting", because I honestly believe that after you've bought your farm and before you've started farming it you enjoy a moment of true solvency. Yes, even if you bought the farm on a payment plan and owe money against it, you're still solvent. Solvent because what you owe is balanced against the value of your farm, something real and tangible. But the minute you borrow money for operating expenses you become insolvent because you owe money against something that has no value to anyone else. If you are able to purchase one cow or a small tractor without having to borrow any money you will have saved twice the value of what you bought. This is because if you borrowed the money, it would cost you half again to twice the amount to repay the loan. And if you borrowed the money that first loan would affect every decision you make until you owe no more. You will keep paying in ways you can never imagine, but most telling you will pay with your independence.

Someone three years back thought he might consider borrowing some money to purchase some beef cows. The banker told him,

"Your credit is good and we look favorably on the fact that your equipment and cattle are all free and clear but we'll have to see a farm plan detailing how you'll be fitting these new cattle into your operation before we can lend this money."

"Why" asked the rancher, *"I've been doing this for years. You know my reputation as a stockman. I'm not going to be doing anything unusual with these new animals."*

"I know", answered the banker, *"but we've got an ag man at the main branch who has to look at all these farm loans and he might want to suggest some changes to your operation that would be beneficial."*

"Now why would I let some banker I don't know tell me how to run my cattle on my ranch?" asked the applicant.

"Well, in truth it might be your ranch, but until you pay off this loan the

cattle are rightfully ours and we do have the right to oversee the operation" answered the banker.

That man withdrew his application for the cattle loan and retained his solvency. But more important he was able in three and a half years to acquire the cattle from ranch receipts without having to borrow the money. If he had borrowed the money, he figured he would have had to make payments for 5 years if he was to keep the loan from hurting operations. With annual fees for rewriting the loan added on to the interest, that man would have paid the bank in five years twice what the cattle were worth.

Almost as an afterword I must add here that if you find yourself with a new farm and the necessity of working at a non-farm job to help defray expenses until things get to the scale you deem right **that is just fine**. Try to make that outside job something you are good at and enjoy. And remember that there is nothing demeaning about having to do your farm part-time. Some of our greatest farmers were and are part-time: Thomas Jefferson, Herman Melville, and Wendell Berry to name a few.

You've bought the farm, now be careful and wary of yourself. Doubt your first impulses. Doubt your fears. Allow yourself the luxury of imagining all that your farm might be before you do anything else. Imagine all that it has been, and try to understand why. Your farm is a lifetime adventure.

•

Chapter Six

"No man is born in possession of the art of living, any more than of the art of agriculture: the one requires to be studied as well as the other, and a man can no more expect permanent satisfaction from actions performed at random, than he can expect a good crop from seeds sown without due regard to soil and season...fixing on an end to be gained and then steadily pursuing its attainment."
- J.C. Loudon, 1825

DESIGNING YOUR FARM

What Sort of Farm

Within the previous chapters we spoke of your need to have a clear view of what sort of farming you wanted to do before commencing with the buying process. The subject of "what farming" may need some help. Rather than give a simple answer as to what was meant, let us hazard to give you enough system possibilities to allow you a "sense" that just might help. It's rather like life, there are no road maps because there are an endless supply of variables.

Perhaps some system samples will also show how important it is that you learn some skills and gather some tools.

First off you must determine your self-imposed limitations. Have you decided that the farming will be an adjunct to other work? For example: Do you want to keep your teaching job and farm? Do you need to put important time into caring for other young or infirm family members and still farm?

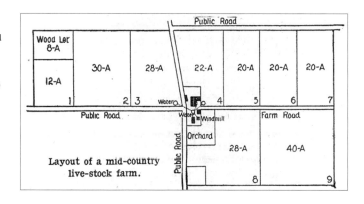
Layout of a mid-country live-stock farm.

"You are no less a farmer just because you have some other interest, endeavor or responsibility. You can do other things also. But you need to keep those things in mind as you design your farm."

Do you want to do art work or crafts and farm? Do you need to work off the farm to help pay the bills and get started?

Let's not think of these as limitations. Let's think of them as realities that need to be plugged into the farm design process. Notice we didn't say you couldn't do these things. Diversity can extend right on into every aspect of your waking day. You are no less a farmer just because you have some other interest, endeavor or responsibility. You can do other things also. But you need to keep those things in mind as you design your farm. Because if you don't give them due consideration early on they will become the seed of frustrations and possibly even failure.

Second, but every bit as important, you must enjoy an honest understanding of WHY YOU want to farm. Is it because it represents a day to day experience, in your mind's eye, that you want for yourself? Is it because you feel driven to be a farmer for altruistic, or "do good" reasons? Are outside forces compelling you to farm? Do you feel pushed? Do you enter the quest certain it will make you rich? Do you want to farm because you believe you can do nothing else well enough? Does the very idea of farming just flat excite you? If you can be honest with yourself about WHY you want to farm it will give you an important framework against which to design the "right" operation. This is so important because it ultimately affects your day-to-day motivations. A farm designed around what you want to do and what you believe you can do will result in a vocation and a life you will be motivated to preserve, build and protect. A farm haphazardly designed without appreciation for your constraints and desires will possibly result in a working environment you lose interest in, one you begrudge, one you feel no compulsion to maintain or build on. An ironic dead-end to what was once an active dream.

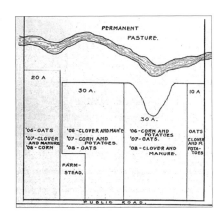

All of this does not subtract from the additional reality that your farming operation may change, even dramatically, after you're down that road. Indeed that is another exciting dimension to this business. It need not be static. Different aspects and features of what you are doing will suggest logical, and illogical, possibilities for change and/or expansion. And the extent to which you carefully plan and design your initial venture will directly affect your continued ability to choose the right direction for later change. And the design of your first venture can and should include a road map of possibilities for expansion.

mapping possibilities
the look and feel of your farm

While with the image of road maps I must suggest that your dreams and your planning may both be helped by the actual physical mapping process. By mapping process I mean literally sitting down with pen in hand and sketching where the house would go, where the chicken house is to be situated and how a cow looks drinking from the trough. Consider doodling and sketching on paper what shape your farm might take. In a recent issue of <u>Small Farmer's Journal</u> I asked children (and then adults) to submit drawings of what their dream (and/or actual) farm looks like if seen by a bird flying over. We received almost a hundred excellent drawings and have published many within the SFJ. Here I want to offer a few to illustrate dramatically how this process can be reinforcement for a dream as well as helpful in planning.

By Sarah Thompson of Florence, Indiana

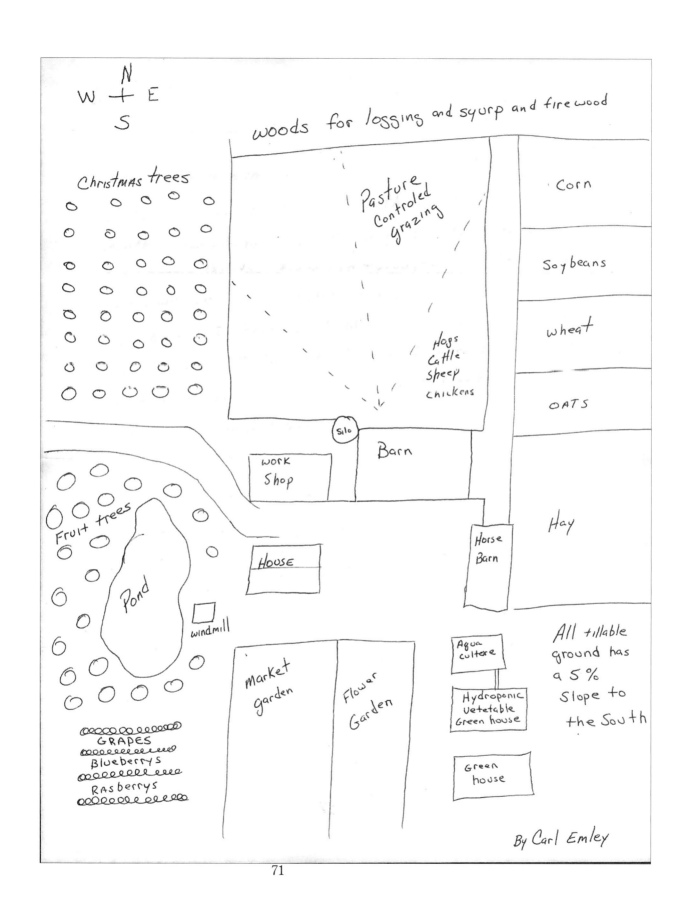

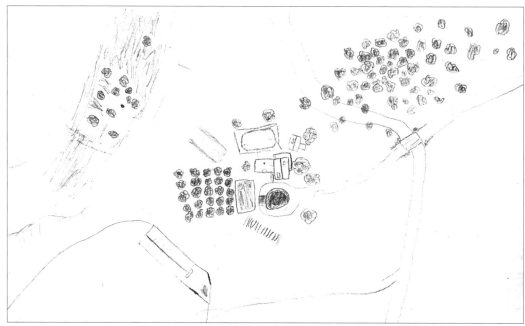
10 year old Donald Gallager's farm has a swimming pool.

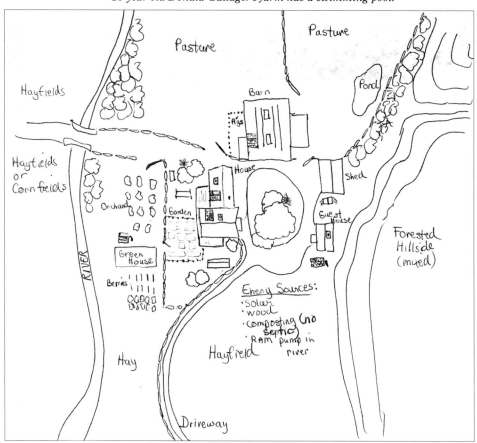
Leigh Gallager, 44, has her dream farm on paper.

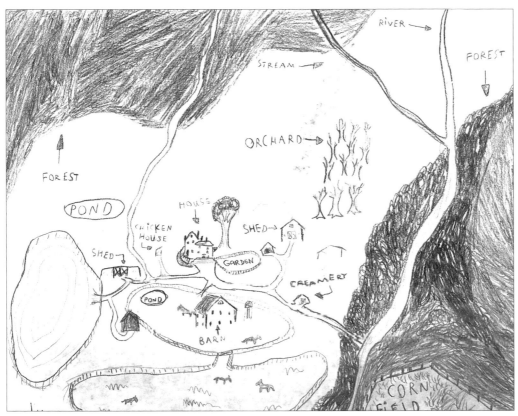
Above 8 year old Rudy Gallager's dream farm

Steve Gallager, 43, drew his dream farm.

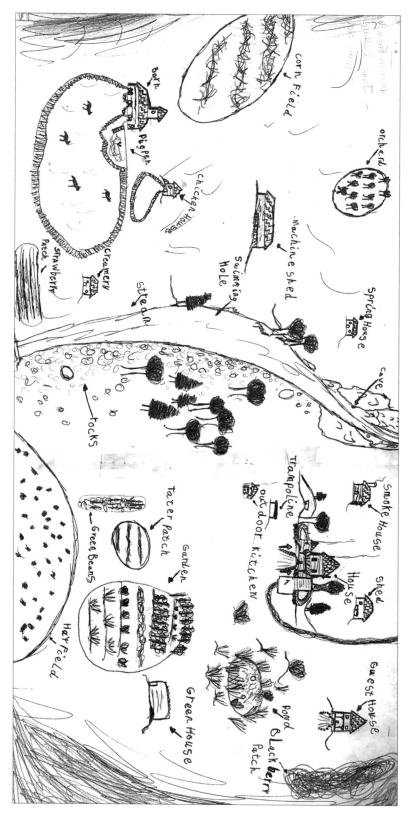

Dream Farm by James Gallager, 8 years old.

System examples

The various components of your planned-for farm can and should have an affect on all other aspects of that farm. Perhaps looking at the promised system examples will illustrate what I mean.

But before we offer up a few working examples we need to knock away at a couple of public misconceptions. The time it takes for a new way of doing things to become tradition has shortened dramatically. Western mass media would have us believe that 2-1/2 to 3 years establishes tradition. And the label "tradition", unfortunately, tends to offer a blanket justification. In this day of the carnival-like ludicrosity of lightning fast technological advances and retreats, mirrored against the decay of the human species and the disintegration of the Earth's biology, we race to create new myths and traditions to fill the gaping hole of our collective uncertainty. Modern science continues to court us in spite of the obvious fact that failings have strained the affair. And one example is the unfortunate trail of industrialized western agriculture. Science and public policy gave us this system which in many cases has destroyed soils, water, communities and farmers. And now, with evidence to the contrary, science would have us believe she can lead us to biological health and new levels of productivity and profit. It won't happen. Not unless we can successfully separate unbiased, unfettered, intelligent inquiry and analysis from the corporate board rooms and the body politic. There is good hope to correct the problems we have but only if we can, collectively redesign what is "do-able" and healthy in agriculture.

Public misconceptions include:

A. Only large scale, well-capitalized industrialized farming ventures can succeed today.

WRONG. Small-scale limited-resource craft-oriented farming succeeds better and is an opportunity not just for individuals but communities and the environment as well.

B. Possible farm system designs are limited to the prevalent examples.
WRONG. And the rebuttal to this misconception is the skeletal concern of this writing.

The seeds of your future successes with farming rest in your ability to always question the norms government, industry and the wider culture would impose on you. Don't just accept what you're told (even here). Always question and measure against WHY you would farm. Hopefully your WHY includes a reverence for life processes, because if it does, your questioning will result in a farming venture which is good for all.

We'll throw out a string of possibilities in three different categories; crops, livestock, and mixed crop and livestock. These examples are abstractions. And the numbers used are purely hypothetical. These examples often mirror working programs, or elements of the programs we've seen across the continent, as well as a sprinkling of our own experiences.

Please note: Small Farmer's Journal has never hidden its philosophical and methodological preference for mixed crop and livestock ventures. But in deference to values and structures outside of our own we strive to offer examples outside of that preference.

Solely for the sake of further humanizing this discussion we've arbitrarily named each example as you might. The names are purely fictitious in this instance and any similarity to real farms is accidental. Each example can be modified to work in just about any region. Certainly some crop types are only suitable in certain climates.

CROPS

A. Kohlrabi Ranch

Forty acres of excellent soils. Two acres in farmstead including kitchen garden. No livestock. Thirty-eight acres divided into six different unfenced 6-acre fields (or areas), with two acres in lane, storage, u-pick restrooms and roadside marketshed.

This example is dependent on value-added direct marketing of organic produce. Heavy dependence on green manure crops and crop rotation. Six acres each year taken out of cash cropping and put into nitrogen fixing legume as green manure cover crop. Over-wintering buckwheat also serves as green manure crop. Note that this system employs no perennial cropping as field areas are in active rotation.

Because of the no livestock parameter, this farm would need a small tractor (30 hp + or all the work could be handled easily with a team). Because of the small acreage of grains grown for human specialty markets (organic cereals and flours) intermediate technologies or subcontracted harvesting are called for. (Insufficient acreage to justify combine investment unless farmer is mechanically inclined and maintains a small older combine. Same applies to other harvest needs.)

$38,000 to $48,000 per year gross receipts reasonable, more is possible.

Work load well spread over year from early spring to late fall/early winter. Peas early harvest (u-pick?). Cabbage and root families (late summer to early winter).

An operation such as this has a steep, critical learning curve with a certain demand for lessons to be taught by fields and the intertwinings, and proximities, of crops. It could take three to five years to perfect the system and enjoy top yields and least cost unless an established successful program is purchased, learned and adhered to. Consider these things if you think you want this sort of farming operation.

B. Spring Creek Apple Farm

Sixty-five acres of rolling hills. Five acres in narrow, eroded, thin-soiled creek bottom. Eight acres in farmstead including packing sheds, cider mill, equipment sheds, greenhouses, kitchen gardens and roadway. Twenty-four acres in mixed woodlot and marginal land including 1-acre pond, 28 acres of established apple orchards including 24 varieties (mostly vintage). Organic orchard practicing biodynamic techniques. Floor of orchard practicing biodynamic techniques. Floor of orchard maintained as clover/grass. For short seasons portions leased for intensive high density grazing to neighboring sheep farmer. Seasonal fish rearing in pond for home use. Pond soil and water carefully extracted for use as fertilizing "manure tea" base.

Additional related ventures. Registered tree farmer practicing sustained yields from mixed age deciduous/conifer woods. Wood high-graded for furniture, musical instruments, and post and beam construction use. Twice yearly home to on-farm biodynamic orcharding workshops. Sales of vintage variety apple scions. Also process apples through cider mill facility and sell through health food store.

This operation is enhanced by diversity and value-added marketing of biodynamic fruit, juice and theories as well as the woods. (Special efforts are made to identify and supply the wood needs of the makers of musical instruments and fine furniture.)

Because of the concentrated orchard harvest workload other projects are carefully planned not to conflict. Seasonal vagaries abound but gross annual receipts could maintain low average of $80 to $100,000.

Building up an operation like this could easily take five years or longer. Orchard requires a substantial initial investment. The purchase of a healthy established orchard should be given top consideration.

LIVESTOCK

C. Footloose Dairy

Seventy-seven acres total. Forty acres in permanent pasture (clover/grass) fenced electrically in 20 two-acre paddocks for intensive rotational grazing. Twenty-four Jersey cows milked same nine months of each year, dry 2.5 to 3 months each year. Calves raised as steers and replacement heifers to sell. Eight acres in farmstead including barns, sheds, lanes, pond, orchard and kitchen garden, 19 acres woods.

Cows fed no supplements, no poison sprays used on

The late Parker Sanborn's Maine Jersey herd

farm. With intensive grazing program one cutting of hay is able to be taken off pastures supplying adequate winter feed for dry herd.

Lynn Miller discing cover crop.

Gross annual receipts $52,000 plus. Opportunities to add value to product and for health food store marketing.

Low cost milk production with added bonus of winter holiday. Classic example of where the net return on smaller operation is greater than net would be on larger one.

D. **Feathered Ranch**

Twenty acres. Ten acres poultry building/run/lanes. Two acres farmstead, 1-acre pond, 6-acres pasture, 1-acre garden.

Raising exotic poultry for all ends of the market: Health food and restaurant broilers, roasters, stewers, and eggs. Cape feathers packaged and sold for fly-tying and ornamental jewelry. Purebred trios sold for breeding purposes. And fertile eggs sold to hatcheries.

Poultry raised include: Speckled Hamburg, Silver Laced Wyandotte, Buff Cochin, Dominque, Blue Andalusian, Partridge Cochin, Mille Fleur, and Araucana Chickens.

Buff and Penciled Runner Ducks, and Black East Indian Ducks.

Pheasants: Golden and Ring neck.

Royal Palm Turkeys.

Ingenious layout facilitates maximizing pasturing of poultry with portable pasture cages for young birds and the careful rotation of a maze of fenced runs allowing that forages are never eaten down completely. Housing is all designed to be cleaned easily and manures are spread over pastures and garden. Single work horse is used for scrapping and pulling one-horse spreader. Runs are constructed of portable netting panels.

Whole grains and supplement ingredients purchased separately and mixed. No growth stimulants or artificial and unnecessary ingredients used. I haven't included any income projections aas there are too many variables.

MIXED CROP & LIVESTOCK

E. Excelsior Stock Farm

One hundred sixty acres mixed crop and livestock, 10 acres maple syrup bushlot, 4 acres farmstead, 2 acres lanes, 1 acre orchard, 1/2 acre pond, 12 acres market garden, 130 acre in rotation as pasture and feed crops. Fifty head Angus beef cows, 20 yearling steers, 5 heifers, 5 grade draft horses, 2 saddle horses, 150 Rambouillet ewes, 500 (2 batches each season) pastured broiler chickens, 12 pastured feeder hogs (in two batches each season) (or approximately 130 animal units).

All work done with horses. Five year rotation includes two year stint in legume grass pasture. Temporary electric fences cordon off 50+ acres of pasture mix into ten 5-acre paddocks, 25+ acres of rye grass into five 5-acre paddocks, 25 acres of oats, wheat, barley mix into five 5-acre paddocks (after harvest), 25+

John Male binding grain in Ontario, Canada.

acres of corn as nurse crop with grass/clover interplanted for weed suppression and pasture head start.

Livestock rotate pasture in progression. In early spring, each group just one day's duration in paddock. Cows, heifers, and ewes get first graze of prime forage, followed by steers, then portable broiler huts, then hogs. In spring total number paddocks 15. Eleven day rest betwixt grazing. In last summer, after grain harvest add five paddocks for one graze before plowing. In fall after corn harvest add five paddocks for one graze before plowing.

Twelve acres of market garden divided between pumpkins, watermelons and potatoes. 80% of all farm family food supplied by farm.

Income sources: Direct sales meat. Live purchases by customers. Animals delivered to custom meat cutter. Premiums paid for unadulterated meats.

Selling maple syrup, beef, lamb, pork, broilers, wool, produce, surplus grains. Possible gross $70,000+ a year.

Intensive full-time operation. Very satisfying to hard-core farmers. Work spread out to facilitate horsepower.

Ohio Amishman hauling pigs to market.

F. **Maine-stay Farm**

Forty-eight acres, 4 acre farmstead, 5 acres strawberries, 8 acres mixed woodlot, 1 acre goose pond, 18 acres rotated (4 year) in alfalfa/clover/grass 2 years; oats 1 year; corn 1 year. Always 12 acres in legume mix divided into twelve 1-acre paddocks. Rotational grazing of 6 milk cows, 3 work horses, 6 hogs, 4 dairy replacement heifers, 3 dairy steers, 25 geese, 500 pullets, 500 layers, 500 chicks (approx. 20 animal units). All feed for animals home raised.

Selling: Direct - raw milk, cheese, eggs, beef, pork, dairy heifers, geese, u-pick strawberries. Possible gross, $24,000 a year.

This example leans heavily on direct marketing off farm. Regular customers come for eggs/dairy products. Special orders taken for holiday geese. Regular customers courted for berries and meat.

Workload could be managed cleverly to allow great chunks of time for other endeavors.

All of these examples should be viewed as nothing but suggestions. Hopefully, you can begin to see the multitude of variables possible and perhaps even enjoy the thought of "designing" your own farm. But more important is the acquired certainty that you must know what you want, and why, before you start. Otherwise success will elude you.

L. Miller imprinting a Belgian foal

Chapter Seven

"Such is the superiority of rural occupations and pleasures, that commerce, large societies, or crowded cities, may be justly reckoned unnatural. Indeed, the very purpose for which we engage in commerce is that we may one day be enabled to retire to the country, where alone we picture to ourselves days of solid satisfaction and undisturbed happiness. It is evident that such sentiments are natural to the human mind."
 - *John Loudon, A Treatise on Forming, Improving and Managing Country Residences, 1806*

GREAT SILLY HOPES

Last chapter we began this discussion attempting to illustrate, through example, what mixtures and makeup might mean to the process of designing a farm. (By "designing" we mean to say "deciding" what a farm will be about as much as what shape it takes.)

An effort was made in that previous segment to state a case for "value" in the designing process. We used the word "why":

Why are you farming?,

Why do you want a farm?,

Why do you want to farm?.

The answers to those questions will all point to "values". They will show what is important to you. A variety of inputs have caused us to want to widen that argument. Yes YOU do need to look carefully at what you value and "WHY" you want to farm. But who are YOU? If YOU are a member of a family who will be involved in this adventure or dream, then every member of your family needs to be involved in the process of identifying what is important. And the reason is that without everyone's involvement the adventure will lack the harmony, balance, diversity, conviction, and strength necessary for success. (This may seem like a repeat of the early chapters but it is too important to your success not to present these concerns once again and in a slightly different way.

WHO AND WHY

Before deciding on what your farm will be about, identify who it will be about. Who it will include? Who are in the nucleus group? Decide what your family makeup is or who all your future family might include? Might your adventure or dream include more than one family? If the farm dream is a solitary one, and you propose to do it entirely alone, it may lack something important. Two or more people working together towards common goals will achieve more than the same number of individuals working independently or at cross purposes. If you are alone but not by choice you may wish to have, or plan to have, help (related or not). That help can and should be included in your nucleus group. When we speak of this "group" we're not limiting ourselves to the "owners of record". We mean to include all those who might have some stake in the success or failure of the venture.

If you are all alone and want to leave the shape of your dream open to include others "when they arrive" always remember that the design may need some reworking in the future to truly include them. And that they need to be a part of that reworking. Don't expect to find partners who will precisely fit the design you work out for yourself. And lest you worry needlessly about compromising your dream, rest assured that you will get back happiness in greater measure than anything you might give up.

So it starts with knowing who's going to

be involved. We're not talking about identifying just those people who completely agree with your dream, your vision of the farm. We're not talking just about those people who would share responsibility for decisions or bills. We're talking about EVERYONE who might be drawn in, affected by, and in the immediate vicinity of the pending adventure. We're talking about everyone who might directly benefit from it's success and everyone who might tamper with its workings. We are talking about your spouse and your children, any other live-in relatives, any working partners, any hired help, any apprentices, all loved ones.

We have received more than one letter at *Small Farmer's Journal* from anxious subscribers saying basically the same thing; "My dream is to have a farm life. My spouse doesn't share this dream. He (or she) is against it. I don't know how to proceed. I'm afraid if I follow my dream I'll lose her (or him). But I know if I don't follow the dream I'll be miserable. She (or he) just doesn't understand."

The above is one good example of why it is so vital to identify the "family" or group who will carry out this dream or project or adventure. And then, and only then, to work carefully and constructively to come up with, TOGETHER, a complete list of what is important to every single person. It should be obvious but it's human not to see it; how can everyone be involved and a part of the process if they are NOT involved and a part of the process? Somehow you need to arrive at goals and a shape to the dream that belongs to everyone involved. You will not be able to succeed if you're determined that everyone else somehow accepts YOUR design for the future of your group.

So together you must come up with a working list of the things you want in your day to day life. Not what you all agree is important, that may come later. First you must just list those basic concerns that, in your day to day life, are most important to each of you individually. And try to distill each item on your list to a positive general statement. It may be a little difficult at first. For example an initial draft of a family's list, including the unlimited contributions of every member, might read something like this;

> *for all of us to work together*
> *a clean, attractive house of our own*
> *nice clothes*
> *honest work*

lots of animals around

a car of my own

peace and quiet

to grow our own food

not to have to worry about the bills

a farm of my own

mom and dad (grandpa and grandma) to be able to live with us

clean water, clean air, clean soil

a tape player of my own

not to be ashamed of having friends over

a tree house

for mom and dad not to have to work so hard

to be able to afford a few nice things

good schooling

peace of mind

not to feel so cramped by all the people around

to be happy

to have a healthy farm with healthy livestock and good crops

not to have to worry about the doctor bills

The second draft of the list, working to distill each line or several lines down to a positive general statement, might read like this;

The whole family working together

A clean, attractive, happy, secure, homeplace

Honest, well rewarded, enjoyable work

The means to afford necessities and some niceties

A peaceful and serene environment

Healthy, homegrown foods, healthy livestock, and a happy healthy family

The ability to save something for a rainy day

A profitable farm for our family and the space it affords us

The means, and the space, to be able to include other generations of our family

A healthy environment in which to live
A way to contribute to the health of the environment
The time to enjoy things, as a family
For our family to take an active part in the schooling of all our members

Nothing has been taken from the original list. Everything is included, in principle, in the second list. The distillation process is important. Now you have the makings of a value statement for your group or family. After you've done this you need to ask every member of the family or group if they have a problem with any statement on the list. If he or she does, then draw a line through the offending statement, and ask everyone for an inoffensive way to restate the original concerns it covered. Your goal is to end up with a list of values that everyone in your group can agree with.

Some families don't need to go to the trouble of doing this in such a literal fashion because they've evolved and grown together so well that they are always cross-checking shared values and shared dreams as well as shared decisions. Yet for others the literal process of getting everybody together and writing these things down has a real positive effect. It helps to realize WHO is really involved. And it helps to show one another that what might have been thought of as opposing values or needs are really quite similar or indeed the same.

HARMONIC BALANCE

All of what has been discussed is with a view towards building an argument for "harmonic balance" as a key towards designing a farm with the best chance for great health, profitability, serenity, sustainability, and beauty.

In alternative agriculture circles it is generally held that an understanding and concern for "biological diversity" must be the cornerstone of any program for truly healthful constructive change in farming systems. And a truly native wide-open unadulterated biological diversity, such as might still exist in some parts of the planet, results in what many refer to as a "fragile" ecological balance. A balance in which every ingredient, no matter how microscopic or intangible, is participant in a living web of interrelationships and interactions. But that living web is multi-dimensional. So to refer to it as having a fragile "balance" is to limit her by description. It is understood that "fragile" is used to denote the modern "environmentalists" perception, born of the incipient arrogance of most scientific inquiry, that nature cannot afford to have any piece of her puzzle messed with. We want to suggest that this logic is flawed. Flawed because it looks at nature as a sum of measurable elements. She is not that. Nature is indivisible and without perimeter or parameter.

Nature works to balance herself within a seamless web of interrelationships and interactions that is our biological universe. But the "balance" is not inherently fragile, it is dynamic and it is harmonic. By dynamic we mean it is strong, and vibrant and constantly changing. And by harmonic we mean that this balance is the result of the harmonizing weave of myriad dimensions of natural activities of every scale. And the weave pulses with ever-changing relationships and rhythms. Many suspect, even fear, the truth that Nature as a whole can and probably will survive mankind. What we mean when we speak of the "fragile ecosystem" is that we humans want things to remain as we (in our limited view) perceive they are today. But Nature is not static. What is both at risk of, and responsible for, the current measure of environmental calamity is mankind's "fragile" tenancy. Understanding the concept of harmonic balance may be a key to giving us some real hope.

We sometimes think of harmonics in terms of music. It is perhaps the easier way to understand what it might be in the picture we strive to paint of the biological universe and of a farm's possible design and health. Play a C note on the piano, and then an E note, and then a G. Now play those three together. Matched they would seem to harmonize pleasantly. Now look at a piano score by Brahms with all those marks on the page. Played by the late Vladamir Horowitz there is a texture, a movement, a flavor, a sound unique to his playing of that even more unique musical score. The identical piece played by the late Glenn Gould takes on a third character. In that original Brahms score the sounds you as

a musician might picture from his writing will always exceed the mathematical sum of their finite parts. The pauses, the phrasing, the tempo, the lingering suggestions that go into the sound you hear, or think you hear, are they all just so much imagining? No, they are real because they are reproducible. And they are dynamic and harmonic because they are changeable, strong, vibrant, and bouncing off one another even as they meld together.

To continue the loose metaphor; think of the Brahms' musical score as a slice of nature or a piece of farm ground. And consider that you and Horowitz and Gould are all farmers. You read the score and marvel at its intricacies and possibilities and genius. Then you work to "interpret" and play the score by first learning the map of its fabric and finally trusting yourself to steward the piece of music through cycles many times over.

Can you see how it is that one could come to a piece of land with similar reverent posture, carefully working to learn as much as possible of the mysteries of her own unique harmonic balance before presuming to farm her?

Now replace the "one", your "one", with the identified family carrying a clear view of its goals and values. Instead of the single virtuoso, you're an orchestra. And instead of the piano score you are faced with a symphonic orchestration. The complexity has grown beyond measure as have the possibilities.

Just as with the human body - we must also be open to the notion that elements of harmony and balance can and do affect the health of a farm and its environs. A healthy farm is a happy farm. And an unhappy farm is never quite as healthy as it might be. And the central nervous system of a farm is its guardian steward family. If the family is unhappy, unidentified, unfocused, uncaring, the farm is weakened.

The successful healthy farm, no matter its component makeup, is a slice of the larger natural world. It too is composed of an enormous web of biological diversity in which interrelationships and interactions are best when allowed, honored, cherished, and wondered after. Everything has an effect and in its turn is also affected. Most powerfully the farm family's outlook, values, and attitude affects the harmony and balance, the very health of their land. And the family is also, in their turn, affected by the good humor of their land.

What we're saying, over and over again, is that your family's tone, attitude and mood are most important even more than the specific farming programs you design because they may carry into your every daily chore a harmonic balance that works its wondrous way deep into the biology of your farm.

Yes you can orchestrate, to a degree, the nature of a given compost by metering the amounts and placement of different organic materials and how the pile is watered and turned. But the compost that results from a great, sometimes comic, calamity of organic wastes, accidentally tumbled together (as with the Nordellian hogs) has a living richness that can only be loved - not measured, not completely understood. Nature takes over with her magic wand creating and allowing catalysts, and chelating, and arguments, and juices, and microbial cycles that struggle constantly for their own balance. Nature, left to her own devises (and our good humor is certainly one of these devises) seeks out correction, balance, harmony, health and that realm of devastation which would fertilize and give birth.

If you would but know who you are and why you farm the resulting humility and great silly hopes you bring to the task would go far to assuring that your farm's true design will have a splendid harmonic balance.

Chapter Eight

"I come now to discourse of the pleasures which accompany the labours of the husbandman and with which I myself am delighted beyond expression. They are pleasures which meet with no obstruction even from old age, and seem to approach nearest to those of true wisdom."

- Cicero, 45 BC

Cost of Starting A Horse-powered Farm

So you have decided to buy a farm and you have a fairly clear fix on what sort of farming you mean to do. Have you given much thought to what will be your power source? Aside from very small hand-labor operations you have three choices: internal combustion / animal power / or a combination of both. You might be surprised to discover that, in the twenty-first century, animal power is a viable motive power option in western civilizations. It is very much so, and I wholeheartedy encourage you to carefully consider it. But then I am prejudiced. I do my farm work with draft horses and have done so for more than 25 years.

But "working horses" isn't for everybody. Some people don't like it or aren't suited and that's okay. Good farming can be accomplished with small tractors when they are used judiciously. And older

Kristi Miller mowing hay with Belgians, Cali and Lana.

tractors and/or horsedrawn equipment can still be purchased at relatively low prices.

This chapter and the next will deal with setting up a horse or mule powered operation. Chapter Ten will touch on the set up of a tractor or mixed power operation. If you aren't interested in true horsepower all I ask is that you take a quick look at the numbers in this chapter before you move on.

Farming on a small diversified scale with horses or mules can be successful and richly rewarding. Those of us who call ourselves horsefarmers know that. And those of us who know too easily come to take our knowledge and experience for granted.

I was reminded of this when taking a recent phone call at the *Small Farmer's Journal* editorial office. A gentleman called requesting information about grain harvesting equipment. He had an 85 acre farm in the East and had spent years at a high pressure job building what he was told would be an adequate savings to move onto his farm and get the necessary livestock and equipment to get started. He had read many books, attended workshops and come to the decision that his was to be a diversified horse-powered farm following biodynamic principles.

This man made the statement that he hoped that $100,000.00 would be adequate to fit the farm with equipment and horses (or, if need be at last resort, a tractor). He had "experts" tell him it would take that much. He was quite surprised when I told him he could outfit his 85 acres with work horses and a complete line of horsedrawn equipment for under $15,000.00.

As part of that same discussion he said he planned for 7 acres of grain every year as part of a rotation. He wanted the crop for livestock feed (including poultry rations). He spoke of how local tractor farmers and repair shops told him he should buy an older self-propelled (or tractor drawn) combine and do custom grain harvest to help pay for it. Another farmer told him that would be a mistake because repairs and upkeep on an old combine would eat him up.

The caller was interested in a horse-drawn binder but concerned about the expense and bother of a threshing machine for just seven acres. That's the point he found himself at when he called.

These last twenty-two years of *Small Farmer's Journal* have been a

marvelous time for the rebirth of work horses, and we like to think that our publication has made some contribution. What was once thought to be a preposterous notion is gaining strong successful converts every day. And we believe that work horses and mules have an important part to play in the making of a new, well-peopled, biologically sound, permanent agrarian system. But only if those of us who believe never stop providing answers to the newcomers - including answers before the questions are even asked.

In keeping with the above, I'd like to share some of that aforementioned discussion expanded to include things we didn't discuss because we ran out of time.

Yes, you can outfit a new or existing small farm with horses and equipment for less than $15,000.00. BUT that does not include the MOST important investment - TIME and CARE in the acquisition of information and experience. I will share some direction with you on how to get the equipment and horses - but getting the experience is harder to direct though it be most critical. A few comments however: reading material is important and _Small Farmer's Journal_ is excellent if for no more reason than it provides verification that tens of thousands do what you want to do. But equally important is the access SFJ provides to those people and the experience, equipment and services they represent. _The WORK_

HORSE HANDBOOK and _TRAINING WORKHORSES / TRAINING TEAMSTERS_ are important references. Reading material is not enough, however. You need to "witness" working examples and "feel" the character of the power sources and "KNOW" how to use THEM. Without these experiences and this skill the relatively low initial investment in animals and equipment would be for naught.

COSTS

How many horses (or mules) you need depends on many variables. For the sake of this discussion let's talk about 4 animals of no particular breed or color, draft in type, 1500 to 1800 lbs ea., well broke to "work" and 7 to 12 years of age. Preferably they all proceed at the same speed when walking with a load. For beginners a few tips:

Allow yourself a year to buy the horses. It may not take that long, perhaps only a weekend, but too many have made mistakes by being hasty with this purchase. Find a good teamster you can trust to help you determine if considered animals are sound and broke. Ask what the horses you're interested in have been used for. If they have been used solely for pulling it might be a good idea to rule those out, at least until you have more experience. The reason is that these horses may be a bit high strung for the beginner. If the horses have been used solely for parades and showing, you may want to rule these out as well. In this case it comes down to whether or not the horses have ever had to really work. If they have never had to pull a heavy load, and warm their shoulders, you may have some real difficulty initially regardless of the fact that they may be well accustomed to carrying a harness and following a bit. All of this is to say that "broke" comes in many sizes and shapes and what you need is horses well experienced with the work you want them to do. If someone tells you that their team is used on the bundle wagon at threshing bees and to spread manure, you have found the ones you want. If however, they tell you this team was only used as a swing team (mid-hitch) for an eight-up...go find a bundle wagon team or a plow team or a corn planting team...Look at the horses feet and ask the owner to pick up a front and a hind foot on each horse. Suspect something if it can't be done. There are so many different things to look for in the general soundness and athletic ability of the horse that, once again, we must encourage you to have some good help available to show you (or explain to you) what you're looking at.

This might come as a surprise: horses of the type you need can be purchased for $700 (low end) to $1200 (high end) each. Surely prices can go higher as auction sales are reporting geldings selling for $10,000 or more! But those geldings will most assuredly be bigger than you need and probably not of the type suitable for hard work. When looking at published sale results from the larger draft horse auctions around the country, it is important to note the total number of animals sold (100, 200, 700!) and that the high sellers commonly represent only about 10% of the total. Horses of the working type we described above always fall into the low price range we've given. Breeders and show people are not interested in the little bay or roan team that will quietly do every job on your small farm.

Here I need to at least tip my hat to those folks who raise top quality work horses worth higher prices yet. There are rare breeds of draft horses such as Suffolks and American Creams and draft ponies such as Norwegian Fjords, Haflingers and others who, along with quality purebred stock from any of the major breeds, are justified in charging considerably more money for their stock.

Truth is a quality purebred mare can do your work for you and pay for herself many times over with the foal crop. But we're talking here about entry level concerns and efforts to hold initial costs down. Time will come soon enough for you to add the extra quality you may hanker for.

As for where to buy: You will be fortunate if you can locate horses for sale on the farm where you have an opportunity to watch them work, and hopefully purchase the harness they are accustomed to. In such a situation you should expect to pay the higher range because it is worth it to you. The lower prices can be found with careful auction shopping. But buyer beware; here is where you really need some experienced help to determine a gem from a big mistake. I am going to hazard a CONSERVATIVE guess that at least 50,000 broke work horses get sold each year in the U.S. For example, in the Wayne - Holmes County area of Ohio 40,000 head of draft and driving horses are sold through auction alone each year according to state figures. Admittedly that figure includes many foals and aged horses and some show horses but consider that this is just a two or three county area! It is easy for the beginner to think that there are very few good work animals available and because of this we need to jump on the first nearly perfect team we find. As you can see from the numbers above, you have MANY to choose from. You owe it to yourself to get what's going to make your farm venture hum. Expect a perfect quiet well-trained team.

As for harness: As noted above try to buy the harness the horses are accustomed to (if it's good) and at least the collars. If that fails, measure the collars the horse's wore to match that size. Used harness of good quality is readily available at the dozens of draft horse and draft horse equipment auctions around the states. But it is mixed in with garbage harness, some of which has been cleverly oiled to make it suggest usability. Get someone to show you what to look for in SAFE used harness.

WARNING: BREAKAGE OF POOR AND/OR ROTTEN HARNESS PARTS IS ONE OF THE MOST COMMON CAUSES OF ACCIDENTS FOR NEW TEAMSTERS!

Used harness can be bought for between $250 to $600 per team set (collars are customarily bought separate). Used collars can be bought for between $20 to $50 each.

But consider this. NEW leather harness for a team can be bought for $900 on up, and Biothane or Nylon harness can be bought from $550 per team on up. New collars can be bought for between $65 to $100 each. In some cases newcomers might wisely elect to go with new harness.

NOW FOR EQUIPMENT

EVENERS. You will need at least one single tree with a grab hook, two narrow double trees and one wide wagon width doubletree.

You will need three neckyokes to match the lengths of the doubletrees you have; you should have one equalizing wheeler doubletree (for a four-up plow hitch), one triple tree (center-fire) and one offset, one four abreast evener that will allow you to simply shackle on your double trees, and swivel shackle grab hook you can button on one of the double trees. Buy this gear used and make darn sure it ain't gonna break soon. All of this at auction can be bought for a high of $300 to a low of $150. Expect to pay more for new gear. (And don't be discouraged if you see some VERY high prices at occasional auctions - that is the nature of the auction beast.)

MANURE SPREADER. Try to get a relatively late model, standard make, spreader (i.e. New Idea, John Deere, Case, Oliver, McCormick Deering) stay away from exotic older makes. You want something you can find parts for. Increase your odds by getting a make that was popular and of which there will be

many "cousins" in hedgerows waiting to donate their organs to your farm's future. There are some good Amish companies doing a great job of restoring mostly New Idea spreaders to NEW condition. Expect to pay a high of $1000 to a low of $200 for a usable spreader. (I have bought them for $25 and $75 ready to go to work - unfortunately I sold them to folks who offered $200 and then I had trouble finding more, only to end up buying replacements for $300 - welcome to the psyche of the scavenger.) New and rebuilt spreaders may go from $1800 to $3500.

FARM WAGON. We're not talking of the romantic wooden wheeled conestoga. Get a rubber tired (or steel wheeled) late model utility running gear (the sort which is still made today) with a wooden flatbed suitable for hauling hay. You can find them from $150 to $800 depending on your taste and luck.

WALKING PLOW. Get someone to help you find this tool. Expect to pay $75 to $500 (the higher price is for a new one).

RIDING PLOW. Again get some help and go for popular makes. The low end should be $250 (but maybe you can do better) and new ones can be bought around $950.

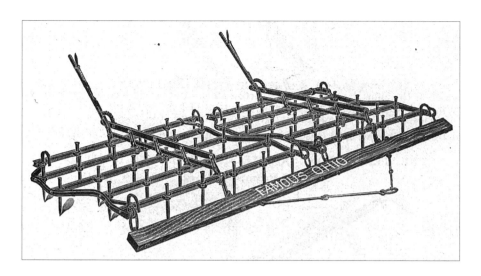

SPIKE TOOTH HARROW. I've had people try to give me these. I know you can buy two four foot sections used for $100 and buy some pretty fancy new ones for $600.

DISC. I don't know of any new discs being made in the U.S. in configurations suitable for horses but there are plenty of used ones. If you are competing with others who know what they're looking at you can expect to pay as much as $600 for top quality. If no one's looking, sometimes you can get a good one for $200. (I bought a great one in 1976 for $40 from a horsefarmer who wanted it to go to a good home.)

ROLLER. Used double roller-packers can be bought at farm sales for $80 to $300.

SPRING-TOOTH HARROWS. Used $100 for two sections. New up to $600.

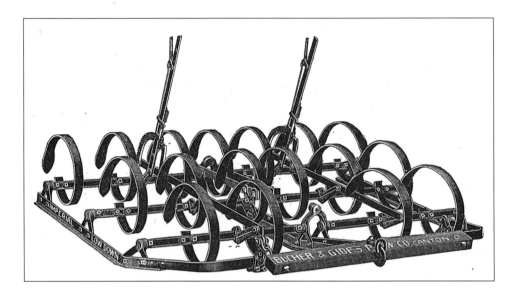

CORN/BEAN PLANTER. Try to stay with McCormick Deering or John Deere. Used $150 to $400.

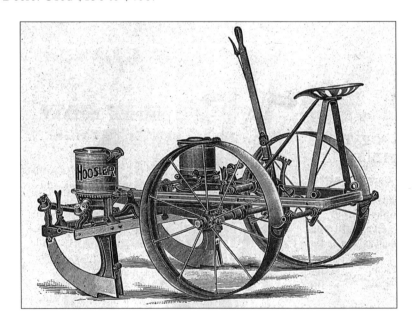

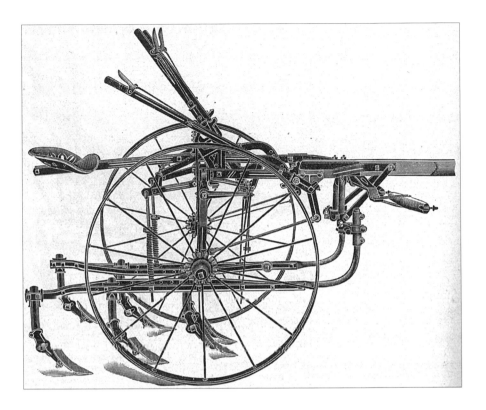

CULTIVATOR. You can get riding straddle row, popular make, cultivators just about everywhere for $50 to $450.

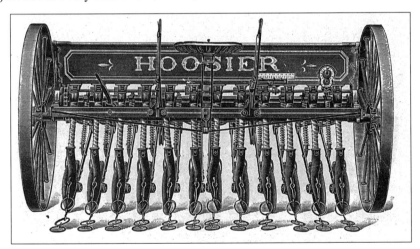

GRAIN DRILL Watch your makes and opt for one with a grass box. $175 low end to $400 high end. A six foot wide team drill would be suitable if your fields are ten acres or less. For larger fields you might want to opt for wider drills. I can drill a ten acre field with a fast walking team in one day with a six foot drill.

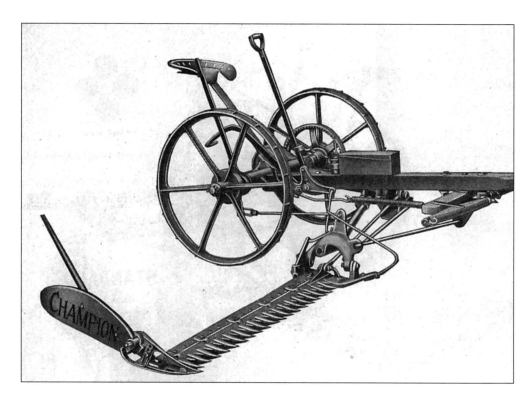

MOWER. Go for J.D. or McDeering (that's shorthand for McCormick Deering which is the same as International). Find help to pick out good ones. Expect to pay $250 to $700. (In 1973 I bought a McD #9 High Gear "yard ornament" for $25 - the guy thought I was crazy for paying that much. I have used that mower every year since then, it has cut well over 1700 acres for me and shows no sign of quitting before I do.) Rebuilt mowers are available from $900 to $2500.

SIDE-DELIVERY RAKE. You have lots of options here as long as the machine is ground driven. With the use of a forecart you can use many modern rakes. For used conventional popular rakes (try to find one that has the reverse tedding position available - they're slick! but not essential) pay $75 to $300. (New rakes have price tags that would surprise a stock broker.)

HAY LOADER. Depends on whether you're going to put up hay loose or bale it. But in the interest of keeping this discussion simple... This tool is highly under-rated. You can sometimes get them for free. $60 to $100 is a fair range.

GRAIN BINDER. Go for a late model enclosed gear (with zerc grease fittings) JD or McD. Sometimes you can find them with a four foot cut but five and six are more common. Yes, a four foot can be operated with two GOOD horses. Find help when shopping for this important tool. Prices vary dramatically according to region. In Amish country a good one will often bring over $1000. Out west in grain country I've seen them sell for $60. I put the range between $250 and $900.

THRESHER. Lots of different makes come into the picture here. Obviously we are talking about a LARGE piece of highly specialized equipment which you may NOT need at all. Price ranges are akin to the binder information above. Keep in mind that you may need to dismantle this to haul it anywhere as they are often quite tall. $50 out west to $1000+ in Ohio.

FORECART. Here I am talking only of the traditional two wheeled and steering three wheeled carts that are most prevalent. They can be built for $50 out of scrap iron or less or bought for up to $800 or more. It may be worthwhile to give careful consideration to the several more advanced designs of forecarts that allow for modern tools to be attached and can range in price from $500 to $2000.

So there you have an admittedly brief and simplistic outline of some of the prices you might expect to pay. Putting them into two columns of high and low ranges and totalling them up you can see we come up with a low investment of $5,915.00 to a high end of $18,300.00. And with a few important variables you can drive those totals up or down.

We have not included everything you would possibly want in your implement shed but you can surely see that starting out on a small farm doesn't have to take $50,000 or $100,000 or more for power and tools. Remember, though, our admonition at the first that without experience and knowledge these numbers won't get you far.

four horses

High end		Low end
4,800	horses	2,800
1,200	harness	700
300	eveners	150
3,500	spreader	200
800	wagon	150
500	walking plow	75
850	riding plow	250
600	spike harrow	100
600	disc	200
300	roller	80
600	spring harrow	100
400	corn planter	150
400	grain drill	175
700	mower	250
300	S.D. rake	75
100	hay loader	60
900	grain binder	250
1,000	thresher	50
250	cultivator	50
800	forecart	50
$18,900		$5,915

There are all sorts of specialized implements such as potato diggers, beet lifters, ground drive sprayers, transplanters, grape hoes, corn binders, balers, self-powered combines, rotary hoes, buck rakes, stackers, graders, ditchers, etc. - and they all have used or junk values of $50 to $1,000 each.

> *On a side note: In that phone conversation as regards the thresher question, I told the gentleman of my experience with storing and feeding Oat grain bundles to horses, hogs and chickens. Carefully stacked in the barn, heads in to the center - butts out - I had no varmint damage. Fed in the manger to working horses I found them to be, by far, the horses preferred forage, and that the working animals retained more digestible nutrients than when whole or rolled grains were fed separate from hay. Hogs did very well on the bundles. And poultry picked the straws clean of grain and hulls, being left with excellent bedding. Following this method I can harvest grain that is a little greener and allow to cure in storage - resulting in more quality in the stalks and perfectly suitable grain. At least in my experience using a binder and feeding bundles, by-passing the thresher altogether, is a viable - if not exciting - option.*

TO CONCLUDE

Now, with social calamities breathing down our collective necks, is the time to get back to basics, foundations, the very structure and fabric of hope. And for many of you hope translates to whether or not you can get on with your farm dream. I sincerely hope that this sort of information will help you to see that your dream is within reach. Because, as I tried to say before, I believe with all my heart that the answer to so many of these problems we are all experiencing is a return to what works, and what is possible, and what makes us part of hope rather than of the hoping. Please never be embarrassed to ask the questions.

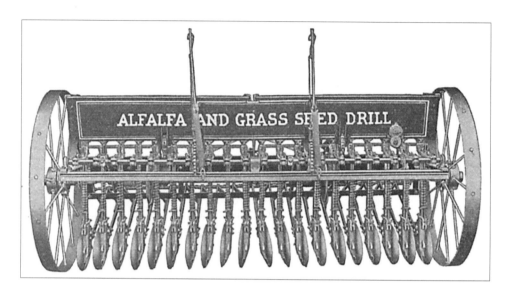

A drill for small seed.

A loose hay loader.

Potato planter

Transplanter

Chapter Nine

"Agriculture was one of the first employments of mankind: it is one of the most innocent and at the same time, the most pleasing and beneficial of any. By its variety, it keeps the mind amused and in spirits: by its exercise and regularity, it conduces to give vigor and health to the body; and in the end, it is productive of every other necessary and convenience of life."
— *Thomas Heyward, 1789*

Setting Up a Horse Powered Farm or Ranch

Part One: Tools

If you are disinclined to use work horses or mules for your farm power source you may wish to skip most of the information in this chapter. I would however, encourage you to read the first two sections "How to Know What You'll Need," and "How to Buy Tools and Implements." There is information there which will be of use to you even if your preference be for tractors.

This and the next chapter are written to offer some assistance to people who are altogether new to the process of setting up and operating farm. Some of the

Bob Oaster raking hay on Singing Horse Ranch

information may seem sophomoric and even silly to those with vast experience in these matters. We deliberately simplify these outlined concerns because we are speaking of beginnings. It is my opinion that a good simple beginning with farming provides the best platform for the gradual successes and challenges which will inevitably lead the intelligent new farmer deeper into the delicious complexities and intricate possibilities of a life with the land. What we're trying to do with this book is give the beginner a somewhat better chance for success. And we do this because we know that the world needs many thousands, if not millions, of new small farmers.

It is possible for someone to successfully acquire a farm at a reasonable price and yet be duped and ruined in the seemingly simple process of getting the tools and supplies to start farming. Although it is a far simpler and more direct procedure to acquire tools and materials for farming than it is to buy the farm, if you don't know what to expect it can be daunting at best and destructive at worst.

You need to know just what tools, implements and supplies are necessary. And you may be benefited in some advice and a few tricks on how to acquire it all.

How to Know What You'll Need

We must assume you have clearly identified what farming mix you will be doing. If you haven't you will not be able to put together an intelligent shopping list of required tools.

It is a mistake to think you will need a basic line of tools which will work for all types of farming. For example: if you're raising poultry or milking cows or raising exotic livestock you <u>may</u> never have a need to plow. So why acquire one?

A great deal of time and money can be saved by acquiring ONLY the tools you will actually use. If you know what type of farming you'll be doing but don't know what tools you'll need, may I suggest that much time and money can be saved if you will seek out at least two other farmers who are already doing what you intend to do. Keep in mind that a next-door neighbor <u>may</u> feel that he doesn't want to help you with information if you are perceived to be a potential competitor. So you may have to go some distance and introduce yourself as a new farmer to someone with a similar operation to yours. Ask them what equipment they use,

An Amishman demonstrates the training level of a team at the Waverly Iowa auction sale.

Bud Evers raking hay at Singing Horse Ranch

and what equipment they feel they lack to do the job properly. It is important to find operations which are of a size similar to your own because the scale of endeavor does, or should, have an affect on the technologies which are appropriate. And when we speak here of size we speak of all aspects; acreage in crop, gross production, available labor, total sales. Will you have more or less people available to work your farm? Will your production need to grow slowly or rapidly? Will the volume of product you have to process, move, store or sell be uncertain or subject to great fluctuations? All of these sorts of questions can and should have a bearing on the determination of just what equipment you will need. For example: will you need a truck and of what size? - can you justify the purchase of tillage implements or should you arrange for rentals, leases or custom farming services? - will you need to consider on-farm refrigeration, drying, or distilling equipment?

It is most unfortunate that I cannot in good conscience recommend that you approach any government or land grant institutions for help regarding tools, supplies and operational procedures. The United States Department of Agriculture has operated from a position of methodological and cultural prejudice for over fifty years. The USDA favors large scale industrialized agriculture and has actively worked to discourage small scale diversified natural farming practices.

A view near Millersburg, Ohio, of 5 Belgians abreast hitched to a pull-type trail plow.

There are notable individual exceptions, however, if you were to approach an extension agent today and request what they refer to as "enterprise data" on the farming you consider you will be discouraged. The information will point in the direction of large scale monocultural operations with exorbitant purchased inputs and limiting assumptions regarding marketing channels. If you were to openly approach an official of any federal farm program bureaucracy with your particular dream for farming 95% of the time you would be ridiculed and discouraged. Please know that there are those who will take strong issue with the above statement and strive to argue that the opposite is true. Remember, you want a small farm - I want you to have a successful small farm - I believe, as you suspect and want to believe, that you can have a successful small farm - I believe the world will be a better place with you on a small farm. Those are your motives and mine. Ask yourself what the motives are of government and industrial officials who would so vehemently argue that you cannot do it and that the world can't afford to let you. Now if those same people become chummy and say they want to help you succeed, WATCH OUT!

The answer; trust your instincts and look to people who are doing what you want to do for information and support.

It is our opinion that, unless you are experienced and well-established, you need to always use caution and start very small in all considerations of equip-

ment. You may even want to modify your plans and wait until your operation demands certain equipment.

the farm shop

There are some basics which come into play - almost all the time. It is the opinion of this author that you cannot farm happily and with appropriate frugality unless you have a well equipped farm shop. This does not mean you need to invest a huge amount of capital in state-of-the-art electric tools. It does mean you should plan on having some covered space where you can do farm repairs and modifications. And you would be wise to accumulate a wide variety of used hand

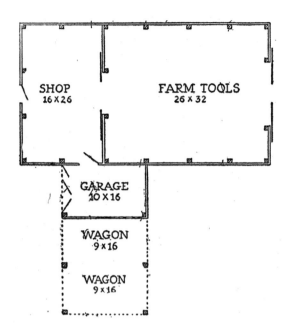

tools including both metric and standard wrenches. Some basic, even crude, metal working equipment can be of tremendous help but only if you have the rudimentary knowledge to be able to use it. Whether you have an arc welder, an oxyacetylene rig, and/or a blacksmith's forge and blower is up to you and your preferences. This equipment need not be expensive. A good heavy duty vise, a basic drill press, and a compressor are examples of tools which can be extremely handy and in some cases downright indispensible. If you are starting out on a tight budget wait and buy cheap (see next section). If you have the luxury of preferences we include some floor plans of possible shop arrangements for you to consider and I am including a list of all the tools we have in our shop as a crude suggestion. Though ours is a horse-powered farm, with electricity, I think you'll find our simple shop equipment will be familiar to most tractor farmers. And for those of you who may be setting out to farm without electricity you'll find some additional notations.

shop and hand tool overview

blacksmith's hammers	sawzall
carpenter's hammers	skill saw
ball peen hammers	chain saw
axes	electrical tester
all types of hand files	propane torch
all sizes allen wrenches	arc welder
hand cross-cut saws	gas welder
hoes	forge & blower
scythes	gear pullers
pitch forks	vises
manure forks	rakes
shovels	leather riveter
peavies	chains & binders
broad ax	ropes
draw knives	jacks
splitting froe	grease guns
bit and brace	bolt cutters
wood bits	pry bars
steel bits	tamping bars
hand drill	post hole diggers
electric drills	all manner of wrenches

tap and die set
router
table saw
misc. carpentry tools
miter saw

chain hoist
anvil
compressor
portable sump pump
fencing tools

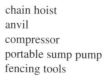
folding draw knife

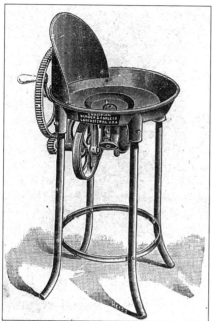
Blacksmith's forge & blower

hand crank seeder

post vise

manure fork
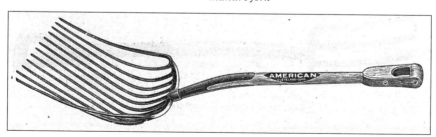

Cradle scythe

Corn sheller

pedal grinding wheel   battery charger

If by preference or necessity you find you will be running your farm without electricity (and it can be done), here is a handy suggestion which is not uncommon on Amish farms and was used frequently in the past. A <u>Line Shaft</u> can be set up to deliver belt power across the full width or depth of your shop. As you can see from the illustration, it is basically an axle with a drive pulley hooked to some stationary power supply (like a small gas engine). Additional pulleys are fastened at various points along the length of the shaft. Belts are fastened and provided

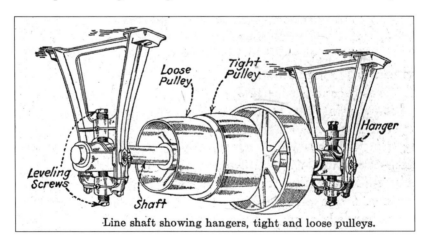

Line shaft showing hangers, tight and loose pulleys.

116

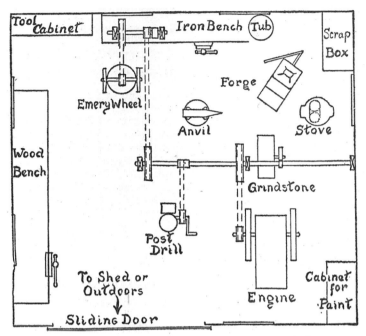
Old-time shop plan employing line shaft

In upper left hand corner of this photo you will see a line shaft setup.

with carrier wheel clutches to allow that different tools be turned on and off. A compressor, grinders, saws, drills and various other tools can be powered quite handily by such an arrangement. Most of industry, from 1915 back, employed line shafts.

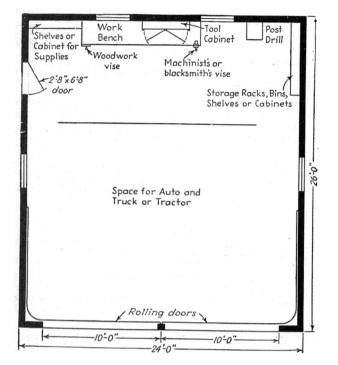

On a visit to an Amish community in the southern U.S. I witnessed the use of an animal powered treadmill setup as the power supply for a line shaft running most aspects of a vegetable processing operation. The power ran a pressure water system, roller tables, and various tools.

At Upper Canada Village, (a living history museum) in Ontario, we watched, fascinated as a flowing creek powered a water wheel which in turn ran a band saw which slowly but accurately cut large logs into beautiful lumber. And outside a Percheron gelding walked in a circle turning a "horsepower" which provided driving force down a shaft to a buzz saw where firewood was being sawn.

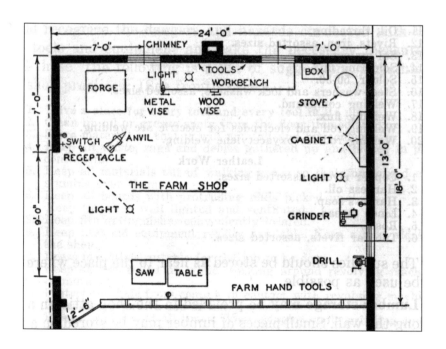

I have not spoken separately of homesteading and self-sufficiency in this text because I view them as central to the successful small farm and a given. Perhaps, however, this is a good spot to comment on some of those items which would be useful or important to the self-sufficient small farm. We, for example, have our own cider press and chest freezer. Others may choose to have canning facilities and equipment along with root cellars and specialized food processing tools including cheese-making equipment. The necessary tools for raising a kitchen garden are a must.

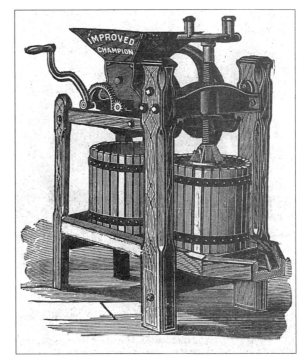

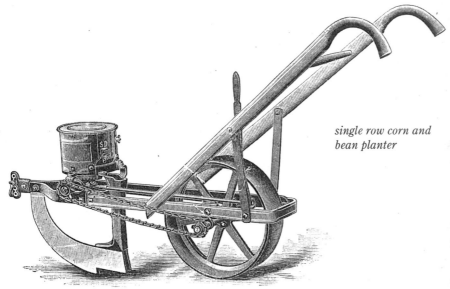

single row corn and bean planter

farm implements

Although we have encouraged you to get information about your own specific farm implement needs and warned that there is no such thing as a list for all farms, I do hazard to provide you with an outline. This should be taken as nothing more than a starting point for some reflection. If you haven't already read the

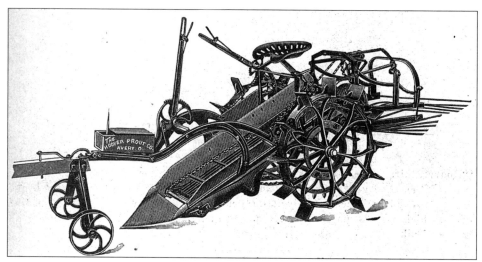
Riding potato digger

previous chapter, you may find some important information there. If you have read it, you may find some repetition or redundancy which is deliberate. NOTE: Please do not go out and purchase all the tools on the list as they may not be what you need.

You should understand that most of the more modern farm implements are designed to be attached and operated by hydraulics and power-take-off shafts which are built-in aspects of modern tractors. In some cases implements are designed to work with just certain tractors and won't "hook up" universally. Many will argue the point with me but I encourage the beginning farmer, whether he be heading for tractors or most certainly if interested in animal power, to limit his search to "pull-type" and "ground-drive" implements. "Pull-type" refers to the fact that the implement usually runs on its own wheels and trailers behind the tractor or horses rather than requiring a fixed attachment to the tractor (as in the case of three point hydraulic hitches). "Ground drive" refers to the fact that any necessary power to operate the implement be provided by engaging the tool's own wheel(s) as a driving force. Most ground drive and pull-type implements are of older manufacture which often translates to them being cheaper.

If all this sounds crazy, confusing, and you feel somewhat lost about this business of starting to farm, relax and take a deep breath. Realizing you don't know what you don't know may be the first step to success. Trust your instincts in this matter. If you feel over your head and inexperienced seek out some guidance, instruction, HELP. The best place to go will be to successful neighboring farmers who are working on a scale and with a style you feel some affinity with. DON'T go to just any old government agent or fertilizer salesman who happens along - they can ruin you in a hurry.

New walking plows at PIONEER EQUIPMENT in Dalton, OH.

Later when you are up and running and about to take your abilities for granted try to remember this moment and offer a young farmer some sound advice. Farmers helping farmers, and would-be farmers, makes for a fertile community.

Farm Implements

(whether it be for a horse-powered or tractor-powered operation)

A rudimentary starting outline

Primary tillage tools	**Secondary**
moldboard plow	disc plow
disc harrow	chisel plow
spring tooth harrow	field cultivator
spike tooth harrow	
roller packer	
row cultivators	

Primary seeders	**secondary**
grain drill w/ grass box	potato planter
corn & bean planter	transplanter
endgate seeder	

Misc

wagon(s)	truck(s)
manure spreader	
fertilizer/lime spreader	

harvest tools

mower	baler
rake	loose hay loader
tedder	buckrake
bale loader	loose hay stacker
potato digger	beet lifter

Besides this rudimentary list there are hundreds of specialized application implements including everything from silage choppers to bean pickers to big rototillers to grain combines. Many of these items are of a scale and cost which may only be justified if you have very large acreage and lots of money. Often smaller appropriate technologies can be found or improvised to accomplish whatever task you think you need to do. Remember that hundreds of millions of individual north american farmers have for 200+ years found ways to get their work done and without big machines. Know that rural back lots are piled high

4 abreast Haflingers pull a motorized baler and two Percherons follow with a bale wagon.

with abandoned farm implements. Also it may interest you to know that older tractors and animal power was, and is, easier on farm tools than the modern behemoth tractors. Add to this the fact that older implements were built to last and you may come to see my suggestion that they may just be better implements.

A little further along I will speak about those equipment items that are specific to animal power.

supplies

Even more than equipment, a supplies list is tied to what sort of farming you will be doing. It is impossible to imagine all that your preferred farming system might require in supplies. We can, however, make an important plug for natural farming systems.

Spike tooth harrow with Percheron team

By my own experience and from the reports of thousands of SFJ readers I can tell you that, without qualification, you can and should farm organically or naturally. This will require an exciting immersion in all that is the craft, and best science, of soil biology and true farming. You will need to learn a great deal but if you enjoy a gripping love of the idea of farming your inroads into the mysteries of plant, soil and animal life will only increase that love.

Simply put, find out what farming inputs are toxic and do everything in your power to stay away from them. Allow yourself to be drawn to possible inter-relationships between all aspects of every living part of your farm and look for ways to use them. When shopping for supplies and the information about supplies you MUST perpetually ask yourself if you understand the motives of those who are selling, or giving. We might reasonably assume that you have some suspicion of farm chemical salesmen (which may, in some cases, be unwarranted) but I want to warn you of the growing number of unscrupulous "organic formula" salesmen. In farming there are no guarantees available for sale whether they be formulas or additives. Look, think and learn. Trust your own instincts. If you have a bug eating your crop you can pay money for a poison to kill the bug - or

A four abreast of Percherons pulling a motorized mower conditioner.

you can physically remove or chase away that bug - or you can feed the soil which feeds the plant in such a way that it is no longer tasty to the bug - or you can forfeit the plant to that bug and move on to other farming.

Supplies may include fertilizers or soil amendments, pest remedies, seed, seed innoculants, livestock feeds, veterinarian supplies, twine, bags, fence posts, fence boards or wire, lumber, fuels, lubricants, watering tanks, and on and on. Some of these supplies should, for safety sake, be purchased new from reputable sources

Organic kitchen garden at Singing Horse Ranch

The author on a buckrake with Anna & Polly.

(which will be around later if you need to "approach" them with questions or problems). But some of these can be either made at home or purchased used. I have successfully saved a great deal of money purchasing fence, building and livestock equipment supplies from farm auctions. In most parts of rural U.S. there are good, to very good, farm supply stores (some of which are coop chain stores) offering almost everything you can imagine. But don't make the mistake of assuming that because your large coop store doesn't carry something it doesn't exist. A little research into the corners of the small farm industry (yes it is an industry) will uncover amazing cottage industries providing access to valuable supply options.

How To Buy Tools & Implements

Might sound funny, suggesting that there is a way to buy the equipment you need. Afterall, don't you just go to the tool and farm implement stores and buy it? Nope. That's a sure way to go broke and spoil a perfectly fine vocational adventure.

Although it doesn't figure into the title of this book, you have no doubt figured out that we are trying to help you get started for as little money as possible. Frugality and thrift are the name of the game here. If you were in the market to buy a motor vehicle and money was tight, would you waltz into the nearest dealership and hand over your wallet? No, you wouldn't. You might go to a used car lot or two and you would doubtless look into all the local ads for people selling their own vehicles. You might even frequent an auction or two. And, if you're patient and smart, you will find what you want and at your price. Buying the tools and implements for your farm should be just such a process.

I am of the personal opinion that nothing beats going to a whole bunch of auctions knowing exactly what you want and being prepared to be as tight as the grain in a good ax handle. It also helps to know a little about how to buy at auction or at least to have an experienced auction buyer help you. An auction purchase is not always a wise purchase. I have seen many folks pay more for an item at auction than they could have spent to buy that same gadget retail. But I have seen far more often folks buy at auction for a fraction of what the item was worth. I'll talk a little more about auctions after this word about budgets.

After you have made yourself a shopping list of what you think you need to outfit your farm, do some additional research. Get yourself a 3 column ledger pad. Write on the left the name of the tool or supply, next to that write down the prevailing retail price, next to that write down a number which is approximately 30% of that retail value, and leave the last column blank. Later, after you have purchased your tool you will fill in the exact dollar amount you paid in that blank column. This will be an important process for you.

Here's a sample of what I mean:

item	retail	goal/budget	paid
four tine pitchfork	$27.95	$10	$7.50
fence stretcher	$24.95	$8	$10
spring tooth harrow 2 sec	$650.00	$220	$45
	budget total	$238	

After you have put together your shopping list, total up the goal column. If the sum exceeds your budget, draw a line through those items on your list which you can wait for. Keep doing this until the sum coincides with your goal. (The 30%

A farm auction at Kidron, Ohio.

calculation is just a guideline, you may instead fill in your estimate of what you think you might pay for each item.) Now as you look for the items on your list ,you can keep track of expenditures and be prepared to make adjustments in your budget. For example, if you've acquired several items at far less than your budget, you might choose to apply that accumulated savings to a particularly fine manure spreader that is a little more expensive than you budgeted for.

With your shopping list/budget firmly in hand, start looking in all the classified ads you can find. You'll seldom see listed the exact items you need. Instead you may read about a farmer's garage or moving sale. If you are able to go and check out the merchandise don't forget your list. And don't forget to tighten up! If you go to a farm sale and find a battery charger for $35 and a side delivery rake for $300, measure that against your budget for those items. Are those prices within the budget? Now here's what I mean by tight: if you have budgeted $40 for the battery charger and $250 for the rake, offer the farmer $20 for the charger. If he says yes, congratulations you've joined the farm economy! If he says he has to have at least $25, offer to buy the rake for $250 if he throws in the battery charger. If he says okay, you've saved $40! If he says he can't do it and is unwilling to come down in price, tell him thanks and move on to the next opportunity.

An Amishman has loaded his auction buys on the wagon for the trip home.

If you want to save money on your budget not only must you buy for less than you budget but you MUST always pay less than is asked <u>even when it is far below your budget</u> with but few exceptions. This will be good practice for you when you finally attend a farm auction.

Buying at auction can be fun, but it can also be terrifying. Here are some tips which might help.

 ಬ It is a kind of financial suicide to decide you MUST HAVE something that is being sold at auction because it means you will probably pay more than its worth and maybe more than you can afford. The contest aspect of the auction can pull you into deep waters. Whatever it is you CAN live without it.

 ಬ Don't be anxious. Watch, listen, and wait. Learn the auctioneer's rythmn and timing. If the auctioneer seems awkward and ill at ease, thank your lucky stars because he or she is probably honest by default. If the auctioneer seems quick, poised, smooth, even slick, prepare to defend yourself even if it means embarrassment. Remember that a tricky auctioneer will use speed and your uncertainty to his advantage. It is always safe and handy to be ready to PRETEND you are confused and misunderstood the bidding. Be ready to state this loudly and clearly as soon as there is any problem. Otherwise that slick auctioneer may have you paying a lot more for the plow than you thought you were bidding.

    ℰ   Don't be the first one to bid on any item. Let someone else open the bidding. Sometimes it will sound as though the auctioneer actually has a bid when in fact he does not. If you want the mower he's auctioning and your budget is $300 wait for someone to open the bidding. If the bidding opens at $300 you are not going to get that mower. If the bidding opens at $75 then goes to $100 you should be prepared to bid $125 and, after they know who you are, slow the bidding down by taking your time to bid. DO NOT EXCEED your budget! In this way you may be able to buy that mower for far less than you budgeted. But remember there will be others, you can live without that mower.

    ℰ   It is a good thing for an auctioneer to take a bid from you and then be rejected by you on the next bid. This helps to establish you as a tightwad who is knowledgeable or at least knows what he wants and his limit on price.

    ℰ   If you are naturally timid or anxious about crowds, auctions may not work for you. However, for many rural folk, auctions are a way of life and you will doubtless find neighbors who relish the "battle" and will consent to bid for you. Just make sure that you write down and repeat the budget you have for the items in question.

    ℰ   Whether at auction or through private sale, it is not uncommon to have opportunities to bid on piles of stuff which might include something you are looking for. Fifteen years ago, I bought at auction six wooden boxes full of old nuts and bolts and misc. hardware. It was about five hundred pounds worth and cost me $35. I chalk it up as one of the best buys I have made. At the bottom of one box I found a set of antique tractor wrenches I sold for $50. Inside another box there was a new/old magneto for a model T which I traded for two doubletrees. The remaining bolts and nuts and washers continue to serve my shop to this day. Sometimes even the five dollar piles are goldmines.

Four Belgians on a modern cultivator.

Part Two: The Special Concerns of Work Horses or Mules

Okay, first off I need to explain that what you're going to read here should not be mistaken for any in-depth complete presentation on what it takes to use horses or mules to farm with. That is a subject for three or four different books. I'm just arrogant enough to think I've written two of them and have my grey matter busy on number three. No, what I'm going to offer you here is just a lazy outline of what you might need to prepare in order to use true horsepower. And part of the beauty of it is that it takes very little.

If you are somewhat uninitiated and approaching the subject of animal-powered agriculture for the first time may I suggest that you consider looking into the reading material recommended at the back of this book.

Do not attempt to work horses or mules in harness without some instruction, guidance and/or assistance. In this instance what you don't know may cause you and your horses great harm.

Very simply; horses and/or mules (and even oxen) can be used to pull implements through the farmer's fields and get the necessary farm work done. Plow a field, they can do it. Spread manure or fertilizers, they can do it. Plant a crop, they can do it. Haul goods, they can do it. They have done it before, as the

Belgian horses pull a forecart and haybine.

country's dominant power source, for a very long time. They have done it all along for growing handfuls of stalwart believers. And they can do it for you today and tomorrow. Broccoli, oats, alfalfa, beans, potatoes, corn (pick your crop) can all be grown successfully by hand, by tractor, by horses or any mix of power supply - the choice is yours. If you choose to consider horses here's some things to think about.

You will need to understand the power needs of your chosen farming scheme and then determine how many horses you need. One horse or mule of moderate size can provide the power needs for the tillage, planting and harvest of up to ten acres (possibly more). Two animals available to work together or as singles provide greater flexibility and can get more ground covered in the same amount of time. Jumping up the scale, 160 acres of mixed farming can be handled by 4 to 6 horses with appropriate tools. Here are answers to frequently asked quetions along with some approximate rules of thumb when computing how many acres.

What is the difference between a draft horse and a saddle horse?

The shape (or conformation) of a horse has a direct relationship to performance at specific jobs. A tall long-legged active horse is better suited for jumping than a short, squat horse would be. Saddle horses (or light horses) have proven to be more suitable, because of conformation, for riding. A large horse with a deep chest, generous sloping shoulder, relatively quiet disposition and large bone is better suited to pulling a load. Draft horses, as a rule, have a slower metabolism and a shorter denser muscle structure. Particular breeds of horses have been developed and refined to perform particular tasks.

What is the working life of a horse?

If begun properly at age 3 or 4, a draft horse kept in good health will work until 15 or 20 years of age and in some instances more. There have been recorded instances of working draft horses at ages between 25 and 30.

How much does a work horse eat?

Allow 1 lb. of grain and 1 1/4 lbs. of hay per 100 lbs. of live weight when performing moderate work. In other words, a 1600 lb. horse would eat 16 lbs. of grain and 24 lbs. of hay. If the horse is not working, hay or pasture alone is sufficient.

How much should I expect to spend for horses?

Grade work horses averaging 1600 lbs. can be purchased from $950 to $2,000. Purebred registered draft horses will range from $2,000 to $20,000.

Is horse drawn machinery and harness still being manufactured?

Yes, harness and hitch gear in ample supply. Some farm machinery is being manufactured in limited quantity. Forecarts, plows, manure spreaders, and tillage tools in good supply. As the demand continues to grow the supply will also grow.

What is the difference between horses and mules?

Mules are a sterile hybred, the result of crossing mares and jacks. The mule has a smaller foot and great heat resistance and is highly regarded for row crop

American Cream horses pull a modern hayrake.

A four bottom tractor plow pulled by 12 head of Belgians.

cultivating in warm climates. The personality and disposition of mules are often the focus of humorous stories. They do differ from horses but are capable of every bit as much work. The only concrete drawback to mules is their sterility.

How young can you start a horse at hard work?

Three years. If you put a younger horse to hard work you stand a good chance of causing permanent bone damage. Occasional light work can be performed as early as two years of age.

Do I need purebred draft horses?

No. Work can be performed satisfactorily with crossbred horses and with light horses (or even ponies). In many instances, the initial investment cost is an important consideration. Many people have chosen to begin with or stick with grade horses to their satisfaction. However, the potential income from the sale of registered offspring is considerable enough to warrant a serious look at the purchase of registered stock.

Can producing mares work?

Yes. For the general health of the mare, it is important to keep up muscle tone. There is no better way than with regular medium duty work. A mare can

Jess Ross on buckrake at Singing Horse Ranch.

work up to within 5 to 1 days before foaling and as soon as 5 days after with no harm to the mare. Care should be taken that feed is adequate and that the mare is not asked to pull hard under slippery conditions.

I have no experience working horses. What is the best way to get started?

Read everything you can get your hands on (including the <u>*Small Farmer's Journa*</u>l, the <u>*Work Horse Handbook,*</u> *and* <u>*Training Workhorses Training Teamsters*</u> available through the <u>*Small Farmer's Journal*</u>). Find local people who use horses regularly and spend time with them. Choose a neighbor you trust as a tutor. Attend any clinics that you can (such as the Work Horse Workshops advertised in the Small Farmer's Journal). And put a premium on experience and docility in the first horses you purchase.

How big a farm can I manage with horses?

One person with four head of work horses should be able to handle an 80 to 160 acre mixed crop and livestock farm. That is to say that there would never be a time when the entire acreage had to be tilled in one season. With additional horses and labor, horses could provide power for up to 600 acres (or more).

How many acres per day can I cover with horses?

The two-horse team will plow two acres a day or disc ten acres. The four-horse team will plow four acres a day or disc 20 acres. The five-horse team will plow five acres a day or disc 25 acres. The six-horse team will plow six acres a day or disc 30 acres. The eight-horse team will plow eight acres a day or disc 40 acres.

What are the drawbacks to using horses?

If you don't like animals or caring for animals, you will find feeding, grooming and harnessing to be tedious. There are no tangible drawbacks for anyone who enjoys horses and hard work. The working speed of horses can often be an asset rather than a drawback.

If I use horses, instead of tractors, to farm, what sort of income should I expect?

You will probably be farming fewer acres but with substantially fewer purchased inputs (i.e., fuel, repairs). With the income from colts and the fertilizer value of manure considered, a competent horse farmer can expect to equal the net income of a tractor farm of twice the size.

Can you make good money horse-logging?

You won't get rich. But a hard-working competent horse-logger in the right area (general economy notwithstanding) can make a good living.

Two Belgians pull a four wheeled motorized forecart and an Italian made rototiller.

Are horses coming back?

Yes, dramatically so. The general feeling that we need to be more in touch with what we do and the exorbitant cost of equipment for logging and farming have combined to account in part for the dramatic increase in the use of horses and mules in harness. It's a trend that started in the early 1970's and it looks to be the sort which will be with us for a long time.

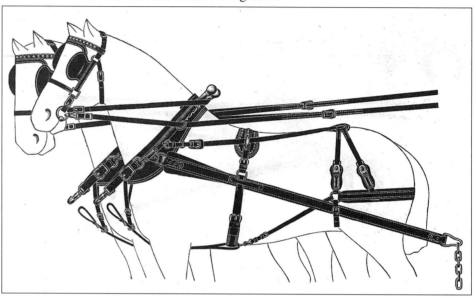

Western Brichen style team work harness

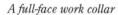

A full-face work collar

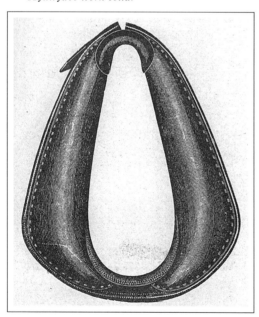

A note of personal philosophy on the subject; we raise draft horses and train them and enjoy having more animals than we need for the work requirements. This gives us the edge when unexpected injury or infirmity should occur.

We have already done some introductory notes on general farm implements. Here I will add some equipment information which is specific to working horses.

Tack

Each horse in the regular work string should have his or her own halter, harness and collar. We keep a couple sets of older harnesses for trainees. The harness for your horses needs to be strong and of the proper size for the animal with collar fit being the most critical concern. (Harness and harnessing is a complex subject so we refer you to the Work Horse Handbook for detailed information.) Included with the harness we keep a range of attachments and adjustable parts which allow us to hook and drive any combination of horses. In much the same manner as we've outlined for farm implements you will need to educate yourself about just what you might need.

brush

You will also find it useful to have some basic hoof care tools including a pick, knife, rasp and nippers. Grooming can be handled with a suitable brush and curry comb although a sweat scraper may come in handy.

Curry comb

It might be useful to keep a horse blanket or two in case an emergency should arise. And last but not least you would be wise in consulting the veterinarian of your choice about what emergency supplies he or she feels you should have on hand. We keep quite a bit of stuff but I might recommend that you should have a rectal thermometer, Banamine paste and Butte tablets at the very least. Your vet should be the one to tell you why.

With all this tack stuff the big investment is in the harness. You have a choice these days of leather harness or a synthetic biothane harness. New leather harness can be expensive but with a minimum of care it is equipment which will last a lifetime or more. Biothane harness is entirely acceptable and costs considerably less. Good used leather harness will usually cost as much as Biothane however the "patient tightwad" will often land good value for small dollars.

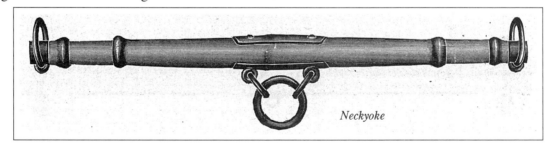
Neckyoke

hitch gear and forecarts

You will need to have a basic supply of neckyokes, single trees and doubletrees. (If you are using only one horse two or three good single trees will suffice.) If you are working larger field hitches you will also need multiple hitch eveners (see the <u>Work Horse Handbook</u> and chapter eight of this book).

You will find it useful to have a variety of clevises or shackles for attaching eveners. One or two swivel clevises will be helpful for certain implement attachments.

As was mentioned before, the forecart allows you to pull draw-type tractor implements with horses. On some farms it is indispensible, on others rarely used. It will be a matter of personal preference. Certainly the type of implements you acquire will have bearing on the forecart's value to you.

The author drives four Belgians on a Pioneer forecart hooked to a pasture harrow.

facilities

Work horses or mules do not require fancy housing. Most farmers I know would love to have a fine, spacious and well appointed stable but few do. Adequate air circulation (or ventilation), some shelter from weather, and a clean manger (which does not trap large quantities of dust) can all be accomplished with a tie-stall arrangement in an open shed. A tie-stall space five feet wide by 8 to 10 feet deep is more than adequate for each individual animal. I personally prefer a double tie-stall which allows me to be able to drive a team into the stall or back it out. Box, or completely enclosed square stalls are okay but you lose the training advantages of the animal needing to stand patiently, back out, and accept

HORSE STALL

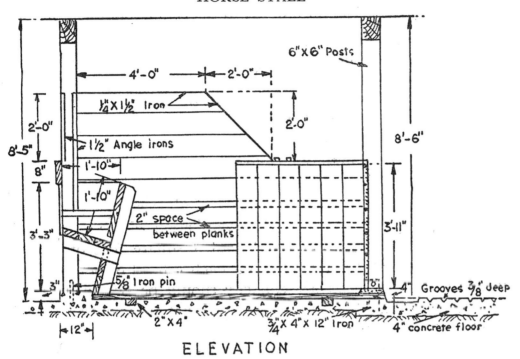

ELEVATION

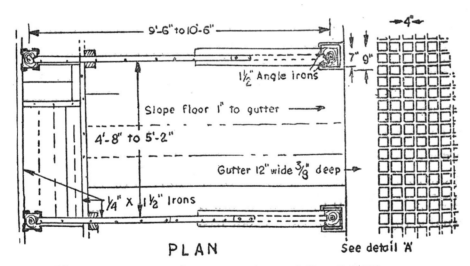

PLAN

(*See opposite page for perspective and Detail "A"*)

being fastened. (see *Training Workhorses*). (If you are unfamiliar with your animals, and they possibly unfamiliar with each other, do not put them together into a double tie-stall without some experienced help. Kicking might result and you or they could get hurt.)

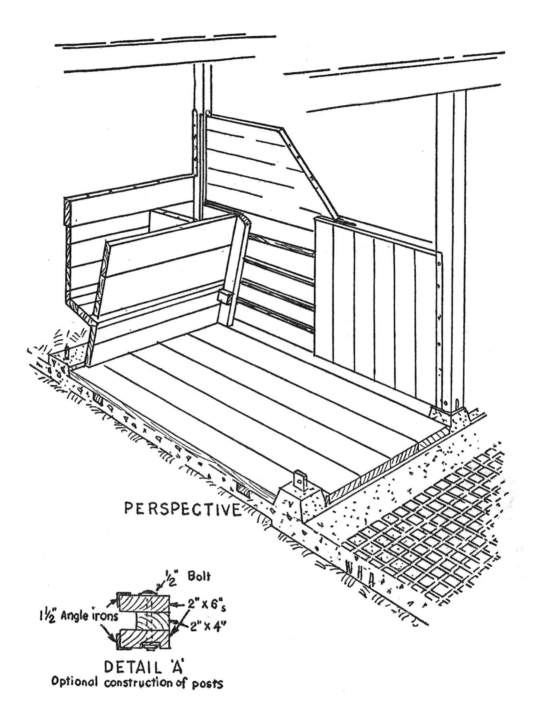

PERSPECTIVE

DETAIL 'A'
Optional construction of posts

You will need a place to store your harness and tack near where you are stabling your animals. You should build a secure space within which you can safely store feed supplements and veterinary supplies. Make it impossible for the animals to gain access to grains and supplements or they may injure themselves seriously. However you provide it, you will need access to large amounts of fresh water for your animals. When not working, your animals will appreciate being able to move around loose in pens or pastures.

procedural effect on layout

When you purchase a farm you may inherit a well established field and building layout. As we said before don't be in a hurry to change this, you need to learn why the farm is set up as it is before making big alterations. But it is entirely possible that you have purchased, or will purchase, a farm or ranch which has a less well established layout or design. You may even end up with a place which contains some of the old original farmstead but with all the land lumped into one unfenced field.* If this is the case and you find yourself having to design a field partition plan, and locate or position future sheds and buildings, take care to include a concern for how it is that you will be working this farm. For example, if you will be working teams of horses or mules you will want to think about being able to access every field at any time in any year. If you design yourself several land-locked fields you will quickly see the problem. While it may be true that our four legged friends will not require the prepared or solid road beds which motor vehicles do, repeated passage over the same piece of wet dirt lane will make a miserable mess. I often find myself taking different routes to the same field work site for this reason.

With experience you will come to understand what I mean when I say that there is no such thing as a universally ideal farm layout. The lay of the land, water and drainage concerns, existing woodlots, roads and other factors all can and should have an affect on design. Agribusiness concerns bulldoze for the largest possible fields, period. Small farmers have the opportunity to consider and respect all the physical aspects of a given site and enjoy the advantages of windbreaks, wildlife shelters, views, and organizational proximity. If all things were equal it would be grand to have each farmstead centered with the fields

---

All across America we see huge farm holdings which are the result of agribusiness consolidation. As corporate farms and large family farms got bigger they would buy neighboring small farms and rip out the fences and any buildings considered less than useful for their operations. Many a chicken house, smoke house, or even corral was bulldozed tragically.

surrounding for easy access to each from that center without a great deal of land going into lanes. However if you must incorporate a lane or two (or more) think about making them wider than necessary and viewing them as occasional long skinny pastures.

Closer to home, if you are going to layout some pens at the farmstead, to facilitate the housing of workhorses, you will save yourself many hours if you can provide for some fenced connection to your barn or shed. Horses or mules which are fed each work day in the barn will quickly learn the routine. If you can simply open gates to give them access you will find they will learn to go to their own stalls to wait for haltering. If instead you must walk any distance and individually halter each animal and lead him or her to its stall minutes will become hours. For example if it takes 20 minutes each morning to bring in the animals by leading them and 5 minutes by opening the gates you can see a 15 minute savings per work day. If you do this for 90 days straight you will save 22.5 hours! What is a whole day or more a year worth to you? The other side of the coin is the argument that if the horses are easy to use you will use them, if they are hard to use you will find some easier power source. Much can be done to make the day to day minutiae of working the horses a convenient even pleasurable procedure. Think through the routines and see if you can't plan some layout features to facilitate your needs.

Chapter Ten

Setting Up a Tractor-powered Farm

Here we will discuss, however briefly, aspects of tractor and mixed power farms which might differ significantly from horse-powered. In the previous chapter we gave outline to many equipment concerns which fit the subject of small-scale tractor-powered farms. I have some experience with tractor farming but as you no doubt have already picked up from this book, I am a horsefarmer by choice. I will try to avoid peppering this chapter with negative comments about tractors and struggle to offer you some

Model C Allis Chalmers Tractor plowing on the contour.

Both pictures are of Model B Allis Chalmers Tractor with two-way plow.

Bedding with two-row bedder and WC Allis Chalmers tractor.

information which might be useful. Please forgive me my prejudice.

As we've said before we encourage you to farm small and with due frugality of applied means. Translated that means be cheap, be sensible, be crafty, be an artisan, and keep things at a scale you can get your arms around (figuratively speaking). Horses tend to dictate such an approach. With tractors it is far too easy to get beyond yourself. For this reason I encourage you, in the beginning at least, to stay away from new or newer tractors. You will enjoy the added advantage of less capital outlay to get started. I had an old friend who once told me that if he couldn't get his work done with his one 30 hp tractor he figured his farm was too big. And that is the big question here; "how many tractors of what size will you need?" The answer, coming from me, is ONE and the size should be compatible with the implements you own (if you already have some). Smaller pull-type and ground-drive tools can always be pulled by a bigger tractor but the opposite is seldom true. If your tractor is too small or improperly equipped with hitch classification, PTO and hydraulics you won't be able to use many newer fancy implements. Stay with the smaller tractor and make your implements fit her. Because there are so many variables once again this will

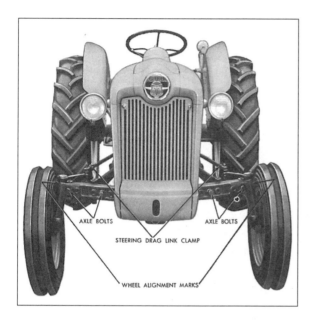
Ford series 800

necessitate some specific research on your part.

The reasons I gave last chapter for going to auctions certainly apply here to tractor shopping as well. BUT if you are no mechanic you will need to have a qualified friend look over any auction consigned tractor before you bid. Major overhaul work on ill-purchased tractors can put a tremendous dent in your budget. A bit of caution and preparation could save you hundreds and hundreds of dollars.

During the middle of the twentieth century there were tens of millions of farmers in North America and many of them had small tractors. Millions of those old tractors are still around and many of them are

Blank listing with B Tractor and single row pick-up lister.

quite serviceable. Whether it is a Farmall M, or a John Deere A or B, or a Ford 8N or 9N, or dozens of other makes and models, patience and diligence will reward you with a tractor at a very reasonable price. What is a reasonable price? That varies from region to region and farmer to farmer. I think anything over

Model C Tractor and two-row lister-bedder.

AC Model WC tractor and tiller disc plow with 6 discs.

$2,000 is a bit much but many knowledgeable farmers will scoff and argue at that. If I was prepared to spend $3,000 each I could purchase several dozen tractors within an hour just in my little county. Where you shop can have a great deal of bearing on supply and demand. Go to areas where there are lots of farmers rather than suburbia.

As a set of beginning suggestions I am including illustrations of a few makes and models of tractors which might be suited to your needs. Some are older than others, some are special application, and some were more popular. It is impossible to illustrate all the suitable choices as there were (and are) hundreds of makes. Might I suggest if you find a lot of a certain make of tractor in your area that you consider going in that direction. It is not an indication that the Allis Chalmers, or the Massey Harris, or the Case, or the Ford, or the John Deere were better because you see more of them. It means, simply that more of them were sold in your area, perhaps a nearby factory or a better dealership or salesman at one time. The reason you should lean in this direction is that when and if you

John Deere Model "A" Tractor— Flywheel Side

need parts and service you will have far less trouble getting it. If you bought a beautiful old Advance Rumley tractor and hauled it cross country to your farm you might have trouble getting parts or even finding anyone able to work on it.

And if you're interested in old tractors it might amuse you to know that there are associations and publications for enthusiasts and these can be excellent sources for problem solving.

Okay, after you get the tractor and implements what besides the well-equipped farm shop will you need? You will need a place to park the tractor out of bad weather. A lean-to shed will work okay and if money's tight a good weather proof tarp will suffice for a season or two.

Last chapter we talked a little about procedures and the layout of your farm. With your trator and the implements you have scavenged you will need to think about maximum overall travel widths and how you'll be getting to the fields. Most spring work is done when the ground can be soft so you might want to give thought to placing lane access at high spots or along ridges.

John Deere models

fuel storage

It is a fact of mechanical life that your tractor, when field work is under way, will drink a lot of fuel. You may in the beginning be able to get away with hauling five gallon cans of fuel in the back of your pickup truck but this will eat away extra pennies and precious time. It will be well worth your time and money to invest in a used barrel pump and purchase fuel in bulk from the local bulk distributor. This will save you money for the fuel and time for less travel. By constructing a separate and somewhat isolated small shed with a floor at a height close to your truck bed's you can slide the barrel into the shed, lock it up for safety and rest somewhat assured knowing it's away from other

Above: McCormick Deering Model 06.

Right: Oliver 550.

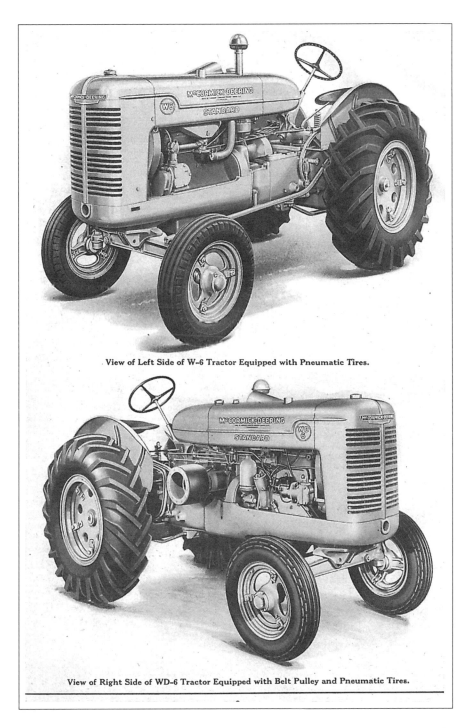

View of Left Side of W-6 Tractor Equipped with Pneumatic Tires.

View of Right Side of WD-6 Tractor Equipped with Belt Pulley and Pneumatic Tires.

flammable buildings. Some will prefer a large elevated fuel storage tank holding 500 to 1,000 gallons and justifying the tanker delivery of fuel. If you're setting up with a limited budget this is an unnecessary initial investment.

Mixed power farming

If you feel that you aren't quite ready to depend on true horsepower but rather like the idea of phasing into it: or if you just prefer to do some of your work with horses and some with tractors here's a few things to think about.

interchangeability: If economy is the name of the game it makes perfect sense for you to acquire a tractor and implements which can be used by either available power source. For example it would be a shame if your plow and disc could not be used by horses or tractor. And it would be expensive to own two different disc harrow designs to facilitate both a tractor and horses. The solution; stick with pull-type implements and stay away from three point hitch

requirements UNLESS you are prepared to go the expense of a suitable forecart with hydraulics and possibly a motor for power take off applications. Such an approach is eventually going to come down to personal preference. My preference is to avoid internal combustion and stick with pull-type, ground-drive, applications. I am not suggesting I am right and the person who goes "modern" is wrong. It really is just a matter of personal preference. What I will say is that a mixed power system dependent on hydraulics, power take off systems and auxillary motors does cost more to set up.

The key "bridge" to allow an interchangeability of implements and procedures between horses and tractors will be the "forecart". Throughout the pages of this book you have seen many examples of forecarts from simple two wheel setups to motorized four wheel machines. You can build your own from scrap iron with some basic understanding of how you need the tool to work. I might suggest that you opt in the beginning for the readily available manufactured forecarts which represent a perfection of balance, strength and adaptability.

stationary power supply: I might further suggest that some of the older model tractors, such as the John Deere model A and B, come equipped with a belt drive wheel which allows that the tractor can be used as a stationary power source to run threshers, silo fillers, buzz saws, saw mills, hammermills, and assorted other specialized tools. In some cases it may be possible to find a belt drive wheel adapter which would allow a power take shaft on the tractor to be converted to this use.

And as a last point I suggest that you make arrangements to house tractors and horses separately, preferably with a little distance betwixt. Some of the fuels and solvents involved with the tractors can be injurious or deadly for horses (i.e. anitfreeze which is poisonous). Also hay, straw, harnesses, and feed grains make for fires which are impossible to put out. Gasoline, Diesel and motor oils coupled with hot manifolds and exhaust pipes provide entirely too much of an explosive hazard to allow close proximity to stabling.

Chapter Eleven

Outbuildings: Sheds & Barns

You may have purchased a farm with a fantastic set of old barns and sheds. You, on the other hand may have purchased raw land with no buildings. Or, somewhere in between you may have land with dilapidated or poor buildings. Any of these situations will require a different attitude combined with your specific operational concerns. I once owned a 77 acre dairy farm with a huge loafing shed, a covered silage bunker, a cinder block milking parlour, and an old original multi-purpose barn in the center. My own farming consisted of a milk cow, feeder pigs, draft horses, chickens, 14 acres of market garden and the balance in hay - grain and pasture. That was more than twenty years and three farms ago but I still remember and am glad that I resisted all temptation to remove any of those buildings. I never made full use of them but others since my departure have. I once lived on a twelve acre farm my parents owned and the only buildings were a 12' x

14' one room cabin and a small barn. On that place I learned the value of every square foot of shed space.

If you know what farming you're going to do then likely you have some clue about what shed or barn space you'll need. If you have some pretty sophisticated older buildings I encourage you not to barge in with major remodeling till you have thought through carefully what you have in mind. The reason is that you might just be building yourself some big regrets in the near future. If you've inherited a fully equipped milking parlor or a meat processing kitchen or a refrigeration unit or a drier or a smoke house or a corn crib - try to use any of these specialized facilities without making substantive changes to their structures or fixtures. You might just find you want to add these endeavors to your farming later or if you should decide to sell this farm find that they add something to the resale value. Once they are gone they are gone. Make sure you need to remove them.

Its beginning to sound like a broken record but you need to know all about your proposed farming operation before you can make intelligent decisions about what specialized buildings you might need. Let's go over a possible basic checklist; besides a house you may choose to need;

        a farm repair shop & tool shed

        a pump or well house

        a wood shed

        a smoke house

        a root cellar

        a chicken house

        portable pig housing

        an implement shed or garage

        a multi-purpose barn

Specialized buildings might include milking parlors, drying sheds, greenhouses, refrigeration buildings, packing sheds, large scale grain storage, large scale poultry facilities, covered concrete composting facililties etc., etc.

If the budget is tight and you are somewhat uncertain about certain aspects of your proposed operation, might I suggest that you consider where buildings might go and what sort of buildings you begin with to reflect a "convenient" design of your farmstead. Think about how you might be doing chores and moving livestock or crops. Can time and energy be saved by a certain organization? Can a building be planned for and only a small portion of it built initially? Can a temporary build-

This magnificent old barn / silo /granary combination exemplifies the height of simple beauty with design following function. The cost of such a building today would be prohibitive for most all small farmers.

These old buildings represent the utility and economy available and necessary for most small farmers. Top; a farrowing shed, middle; a poultry house, and below; a cattle barn.

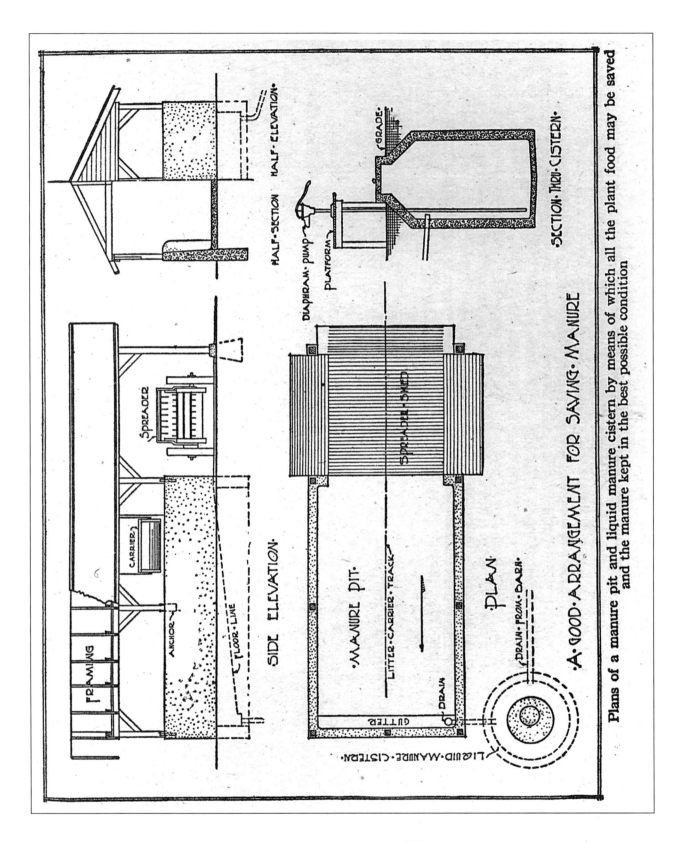

Plans of a manure pit and liquid manure cistern by means of which all the plant food may be saved and the manure kept in the best possible condition

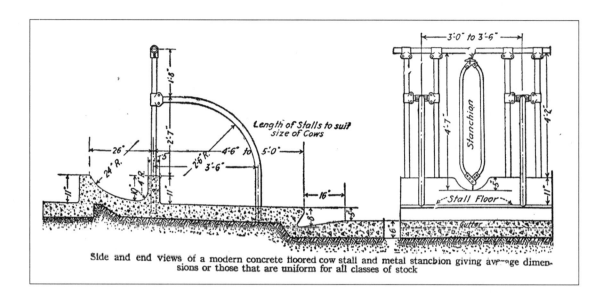

Side and end views of a modern concrete floored cow stall and metal stanchion giving average dimensions or those that are uniform for all classes of stock

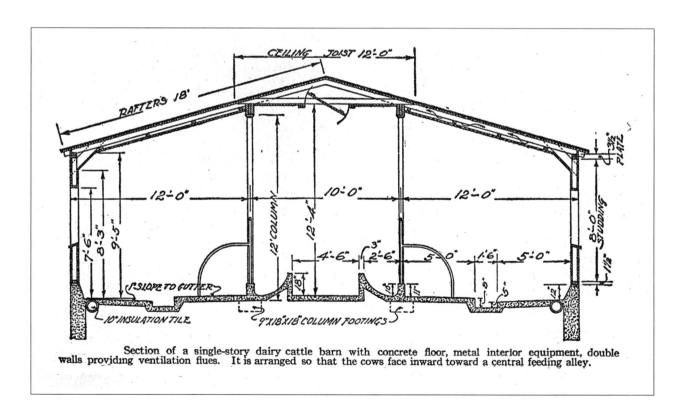

Section of a single-story dairy cattle barn with concrete floor, metal interior equipment, double walls providing ventilation flues. It is arranged so that the cows face inward toward a central feeding alley.

BLUE AND BLACK FEATHER PLATE (FEMALE)
No. 1, Barred Plymouth Rock; No. 2, Blue Andalusian; No. 3, Houdan:
No. 4, Silver Spangled Hamburg.

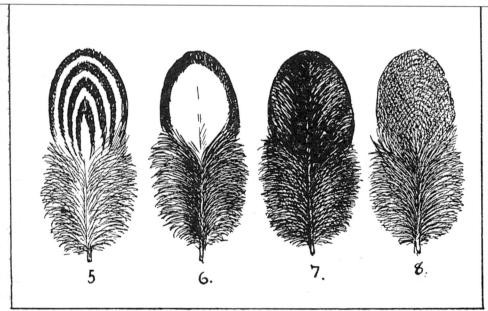

BLACK AND WHITE FEATHER PLATE (FEMALE)
No. 5, Silver Penciled; No. 6, Silver Laced; No. 7, Black; No. 8, Silver Grey or Duck-wing.

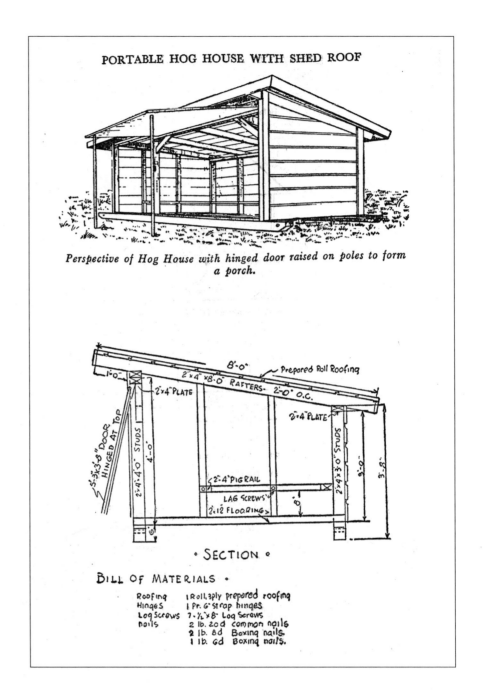

Perspective of Hog House with hinged door raised on poles to form a porch.

ing be designed for a different location and purpose in the future? For example, say you want to have thirty laying hens right away but are thinking about raising from 500 to 1500 chickens in the future: why not design a small hen house on skids built so as to be portable and usable later as a potting shed or incubation room or tool shed or lambing barn? All that would need to be incorporated into your design would be simple factors such as necessary head room, window placement, width of

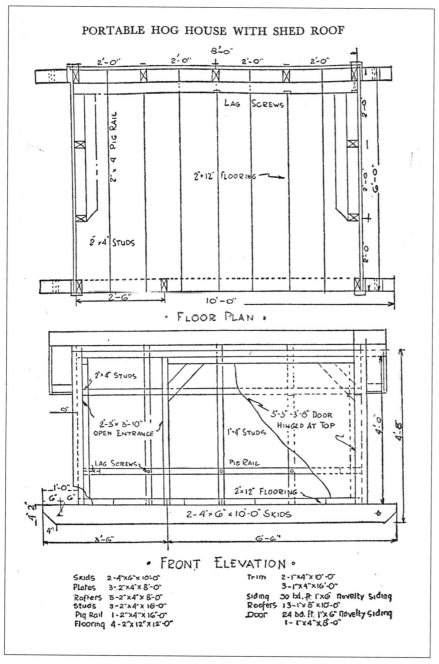

doors and floor strength.

In this chapter I am offering some old, favorite, building plans for simple sheds and barns. Please accept these as they are meant, simply as a beginning suggestion to get the design juices flowing. If you are practicing necessary frugality you will need to build with the materials you can find, barter for, or afford and those materials will suggest certain design possibilities and limitations. For example, if

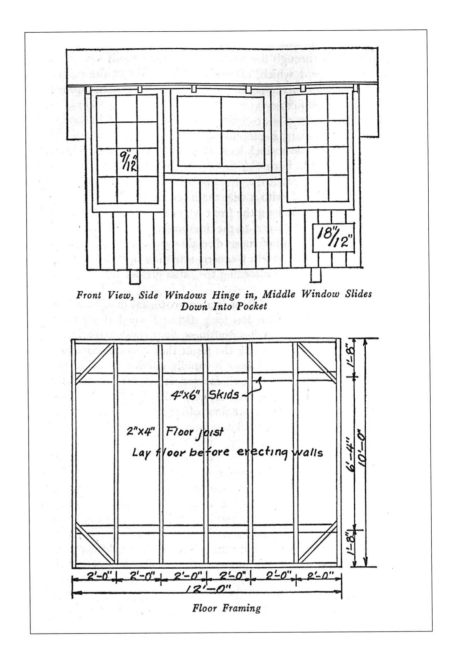

Front View, Side Windows Hinge in, Middle Window Slides Down Into Pocket

Floor Framing

you have lots of free poles or rock, use the stuff. If you have been given a metal clad building if you take it down, use the stuff. You might have some available materials which you find unpleasant to look at, find a way to make them passable or even beautiful and use the stuff.

Fences. Your farm will be viewed as a giant lunch box for predators whether you have a mixed crop and livestock operation, strictly livestock, or just crops. And

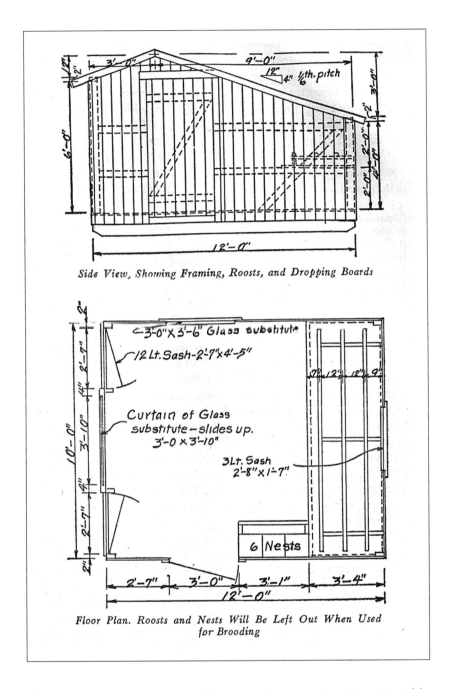

Side View, Showing Framing, Roosts, and Dropping Boards

Floor Plan. Roosts and Nests Will Be Left Out When Used for Brooding

this will be true if you're out in the wilderness, like us, or have lots of neighbors nearby. I suggest you view each new perimeter fencing project with these concerns:

    a. Will I want to discourage dogs and/or coyotes from entering this field?

    b. Will I need to discourage deer, antelope and/or elk from entering this field?

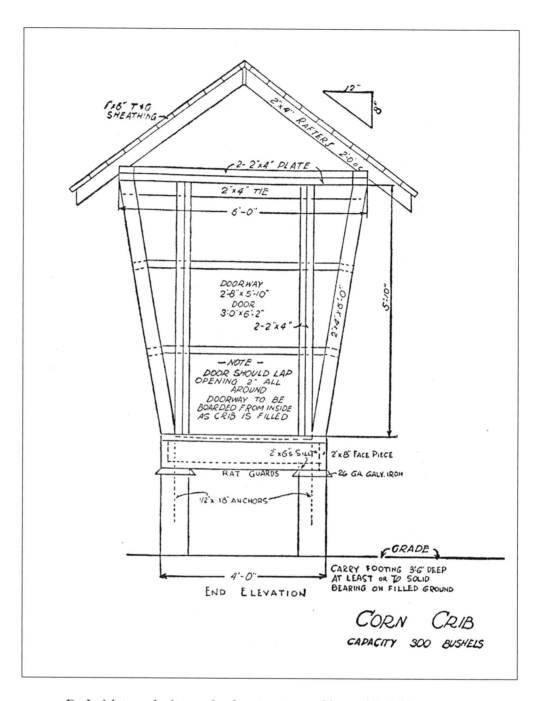

c. Do I wish to make it easy for deer to enter and leave this field?

d. Do I want to provide cover for beneficial predators?

e. Will this fence traverse a line which parallels a need for protection from wind?

When I see a large piece of agricultural land with no trees, no hedgerows, no swales, and no fences it makes me sad. I know it was once farmland but it isn't any

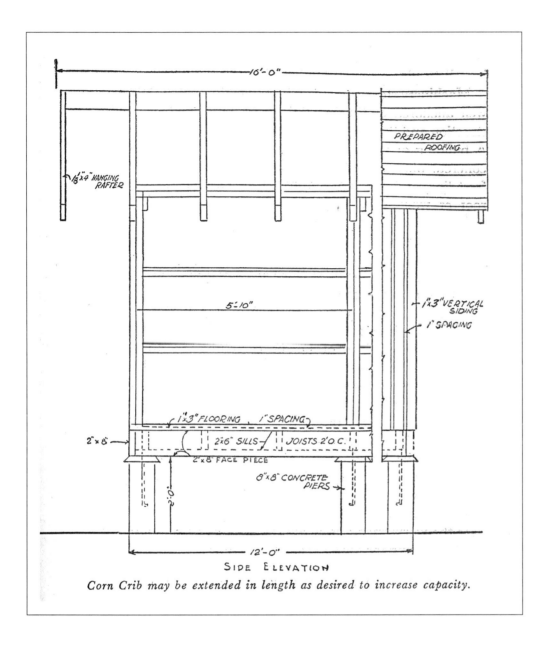

Corn Crib may be extended in length as desired to increase capacity.

longer. Now it's a site for industrial agribusiness nonsense. "We're gonna use this site to grow as much food as we can sell for as much money as we can get, period. You got a problem with that?!" Oh brother, have I got a problem with that. Lots of problems! None the least of which is that this picture I paint is of a sterile landscape: dead soil - no animal life (including humans) - no insect life - only a biomorphic excuse for plant life. And the first thing missing from the picture are the fences. Then go the borders, hedges, margins, trees, bushes, uncut grasses,

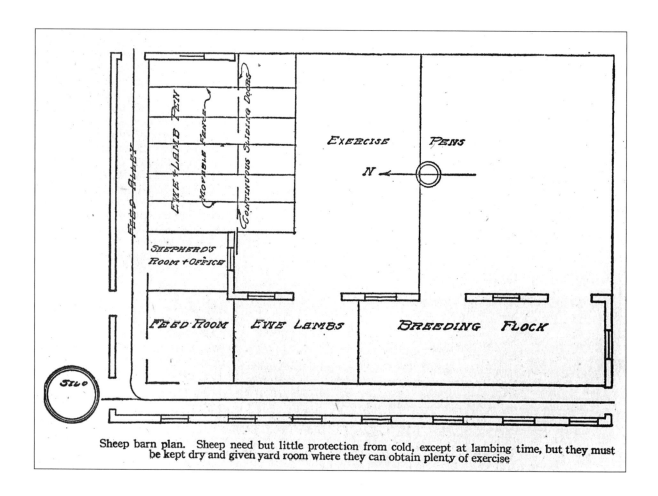

Sheep barn plan. Sheep need but little protection from cold, except at lambing time, but they must be kept dry and given yard room where they can obtain plenty of exercise

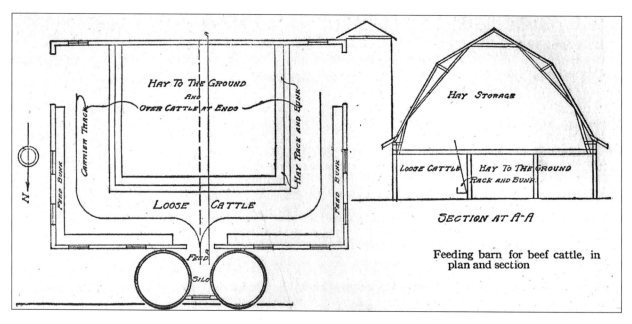

Feeding barn for beef cattle, in plan and section

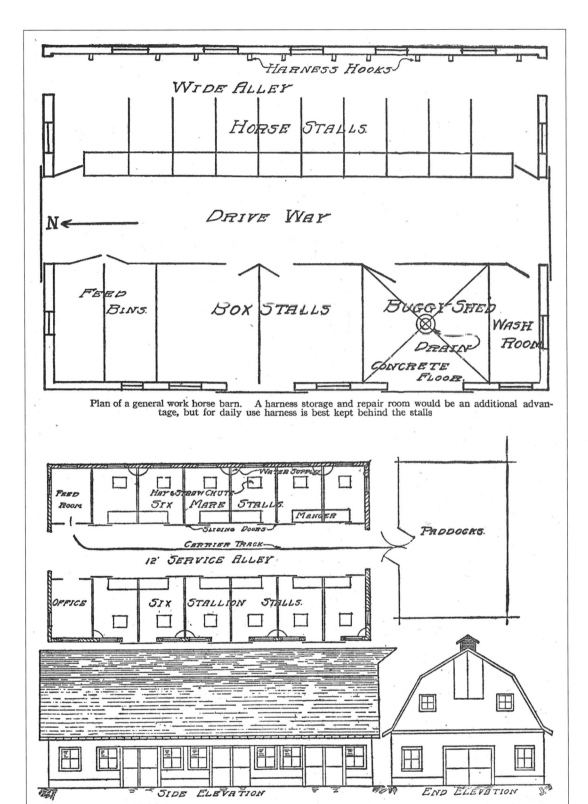

Plan of a general work horse barn. A harness storage and repair room would be an additional advantage, but for daily use harness is best kept behind the stalls

Plan and two elevations of a stallion and mare's barn. The professional breeder must build attractive structures in which to receive buyers as well as efficient ones in which to care for his stock

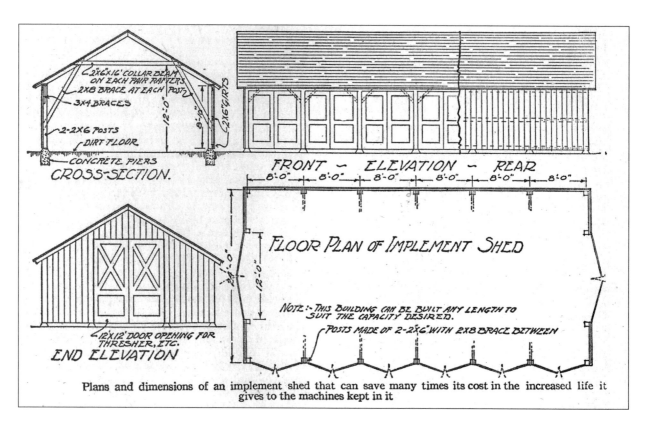

Plans and dimensions of an implement shed that can save many times its cost in the increased life it gives to the machines kept in it

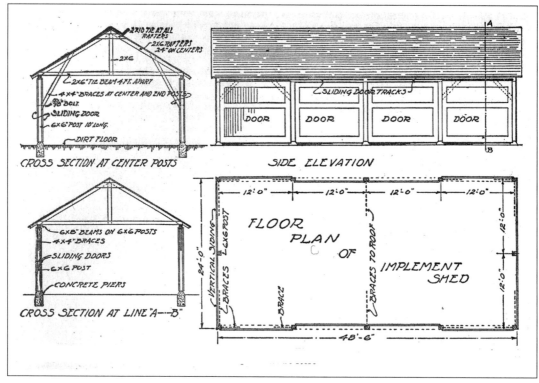

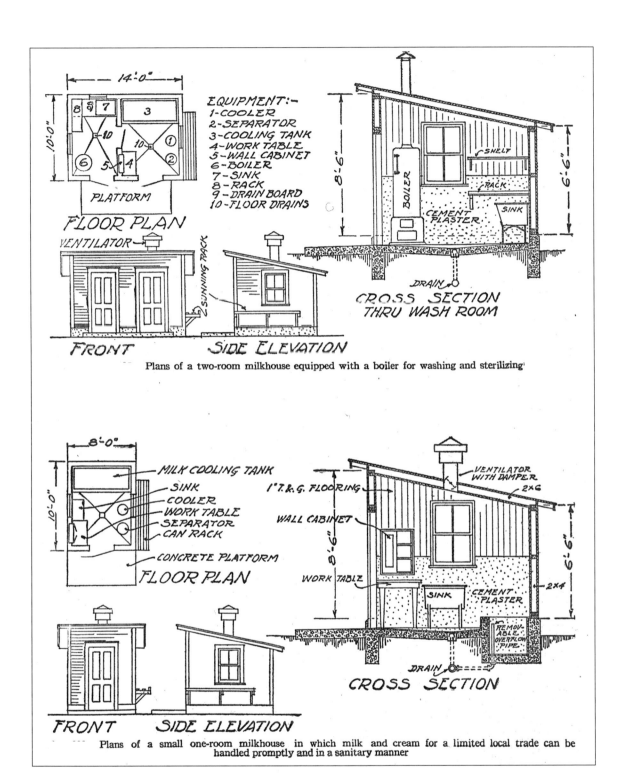

Plans of a two-room milkhouse equipped with a boiler for washing and sterilizing

Plans of a small one-room milkhouse in which milk and cream for a limited local trade can be handled promptly and in a sanitary manner

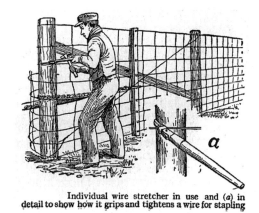

Individual wire stretcher in use and (*a*) in detail to show how it grips and tightens a wire for stapling

hideouts, havens, meeting places for bugs, birds and varmints.

When I think of small farms I see fences, and hedgerows, and trees, and shaded lanes, and grass-bordered ponds. I see a patchwork, the overall design of which speaks to the invisible symmetry of working interrelationships.

When you build a fence, build it with everything in mind. If you want dogs kept out, build it with a barbed or electric wire stretched low to the ground, this will at least slow them. If you want to keep deer and elk out of the field, remember that height may not be the answer. It often works better to have a taller than average fence built leaning into the oncoming deer. If you wish to welcome the deer to cross your fields, think about making any wire fence out of hi-tensile smooth wire which can be built to take an occasional electrical charge for maximum livestock control. If you want to provide cover for animals, think about stone fences, hedgerows and the possibility of brush plantings along a wire fence. I fancy the idea of building triangular enclosures at corners to further support the fence and provide initial protection for new tree plantings. And while you're out there building that fence, perhaps it would be a good time to consider planting a row of

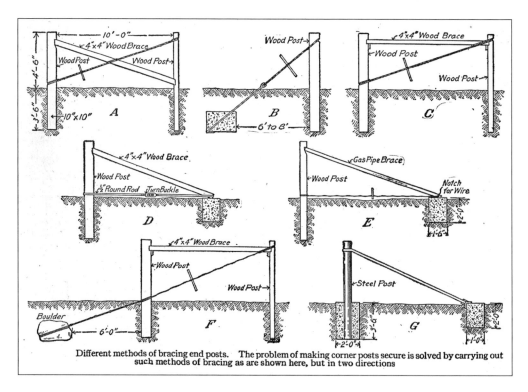

Different methods of bracing end posts. The problem of making corner posts secure is solved by carrying out such methods of bracing as are shown here, but in two directions

fast growing trees for a future windbreak. Each of these things adds to the character and fertility of your farm and serves to potentially influence the character of the surrounding countryside and community.

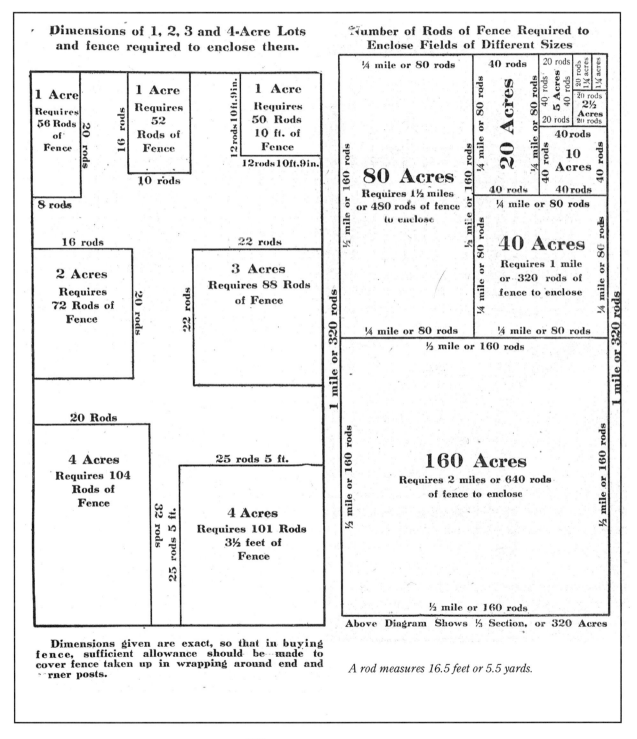

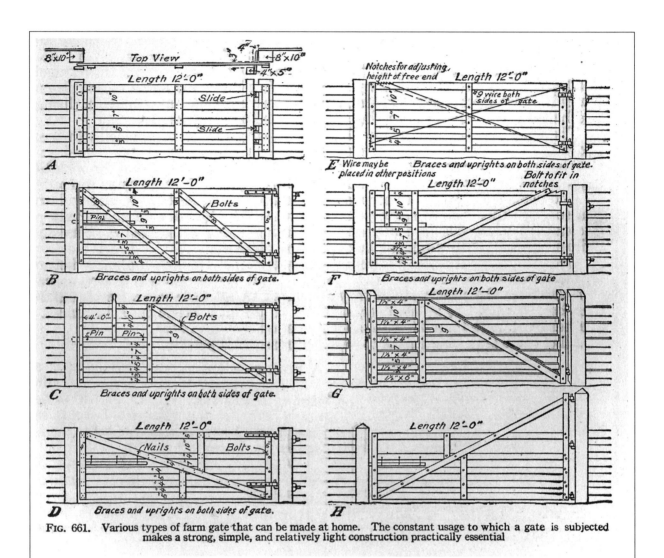

FIG. 661. Various types of farm gate that can be made at home. The constant usage to which a gate is subjected makes a strong, simple, and relatively light construction practically essential

Chapter Twelve

Naming Your Farm or Ranch

Back in the Fall of 1984, *SFJ* reprinted an older article by Englishman K. A. Kirkpatrick on picking a name for your place. He conjectured that we'd all be happier to be identified as the owner of "Clover Crest Farm" or "Glacier Meadow Ranch" rather than as *'those folks living on a farm on Smith Road one mile out of town'*.

Market identification, with packaging clearly marked and signage on the property could indeed have business value, especially if you are interested in U-pick or on farm sales. Kirkpatrick further felt that naming the farm and hanging out an attractive sign pushed the owner to look a little differently at his premises perhaps even resulting in timely sprucing up.

Though some farmers have a legitimate concern that naming their place is foolish and seems vain, many will find the move appropriate and helpful in the overall setup of the farm.

If you decide to name your place, bring the whole family and all the helpers into the task It will give them a pride of involvement and identification that goes back to reinforcing the WHY of the whole adventure. Here are Kirkpatrick's own words in suggesting how to select or design a name:

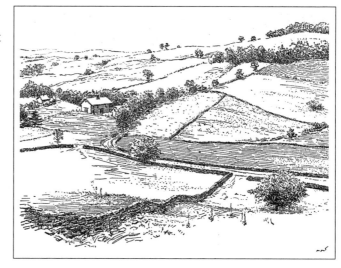

"Preferably the name should have some relation to a distinct feature of the farm. An illusion to family history, an attaching bit of sentiment, a poetic bearing, or a musical sound, is by no means to be condemned; either of them may lend endearment, to supplement that which naturally attaches to one's home. It is questionable whether the advice sometimes given to 'let the name indicate the line of farming pursued' should be followed; for it puts too much accent on the commercial side of farm home life.

Suggestive Suffixes and Root-Words

Following is given a list of common suffixes and root-words used in farm and place names. In another place will be found a suggestive list of combinations. Others will readily occur to the reader:"

Aerie	Mesa	Mere	Plain
Aere	Moor	Nook	Prairie
Aire	Doun	Isle	Point
Aer	Doon	Idyl	Pond
Butte	Eyrie	Knoll	Run
Bluff	Eyry	Knob	Rill
Brook	Field	Kopje	River
Brae	Forest	Kill	Rest
Burn	Ford	Lane	Ridge
Bourn	Glen	Lough	Spring
Bourne	Glade	Lea	Tree
Croft	Grove	Land	Vale
Crag	Glyn	Lawn	View
Craig	Heath	Lodge	Villa
Canyon	Hall	Lake	Vista
Crest	Hill	Loch	Valley
Cliff	Halm	Mount	Wald
Cleve	Home	Manor	Wold
Creek	Haven	Mead	Wild
Dale	Hurst	Mede	Water
Dell	Hollow	Meadow	Wood
Down	Ideal		

Farm and Place Names Derived from
Common Root-Words and Suffixes

Gleneyrie	Cedar Brook	Forest Manor
Glenaerie	Brier Brook	Forest Lodge
Aeriecroft	Forest Hall	Forest View
Airedale	Crystal Brook	Forest Drive
Hightop Butte	Bonny Braes	Forest Hill

Snowy Butte	Fair Brae	Mount Forest
Orchard Butte	Fairburn	Forest Nook
Pine Tree	Butte Clerburn	Forest Croft
Scott's Bluff	Clerbourn	Forest Lane
Smith's Bluff	Claybourn	Forest Mead
Pine Bluff	Woodburn	Forest Mere
Cedar Bluff	Heathburn	Forest Lake
Lindwood Bluff	Hillbourn	Hazel Dell
Orchard Bluff	Forest Home	Fern Dell
Poplar Bluff	Forestholm	Maple Dell
Meadow Bluff	Forest Run	Cedar Brook
Bonny Brook	Forest Springs	Shady Brook
Laughing Brook	Forest Glade	Brookside
Meadowbrook	Monteleve	Glynloch
Brookwald	Cliff Home	Glen Lane
Brookwold	Clifholm	Glenwold
Brookholm	Cliff View	Glen Hurst
Brook Home	Far Cliff	Glenwood
Brookhurst	Cliff-by-the-Lake	Glen Croft
Cedar-Croft	Crooked Creek	Glenwald
Home-Croft	Willow Creek	River Glen
Croft-in-Vale	Smith Creek	Glen Loch
Clovercroft	Maple Creek	Glen Lake
Fieldcroft	Rock Creek	Glen Acres
Cottage-Croft	Shadow Creek	Brood Heath
Crag Home	Plumb Creek	Fair Heath
Craigholm	Lindale	Pleasant Heath
Craigaerie	Avondale	Hilltop
Crag Eyrie	Smith Dale	Orchard Hill
Glen Craig	Maple Dale	Cedar Hill
Craig Place	Cedar Dale	Willow Hill
Hill Crest	Auburn Dale	Poplar Hill
Montcrest	Dale End	Maple Hill
Crest-on-Lea	Swale Dale	Smith Hill
Meadow Crest	Linden Dale	Hillside
Clover Crest	Pleasant Dale	Vine Hill
Briar Crest	Wood Dale	Berry Hill
Briar Cliff	Dale-in-Wood	Guernsey Hill
Montclif	Hazel Dale	Jersey Hill
Rose Dale	Round Hill	Wildholm
Berry Dale	Hill Run	Holmwald
Oak Glen	Hill Lake	Eastholm
Glen Dale	Hill Acres	Southolm
Glen Home	Hickory Hill	Westholm
Smith's Glen	Oak Hill	Northolm
Jersey Dell	Rose Hill	East Home
Clover Dell	Pleasant Hill	South Home
Southdown	Prospect Hill	West Home
Long Glen	Fairy Hill	North Home

Orchard Glen	Bill-o-Hope	Homewood
Glen-o-Mine	Sunny Hill	Sweet Home
Glyn Lea	Shady Hill	Happy Home
Glyn Place	Cherry Hill	Glen Home
Glyn Fair	Linden Hill	Mountain Home
Eden Glen	Holmwood	Orchard Home
Glyn Haven	Fairholm	Fair Home
Glyn Mere	Cedarholm	Helmwold
Mountain Glen	Home Lea	Island Hill
Clover Home	Mount Isle	Orchardlands
Cloverholm	Meadow Isle	Roseland
Feldheim	Cedar Isle	Beautyland
Field Home	Maple Isle	Prairieland
Fair Haven	Belle Isle	Timberland
Deep Haven	Idylwild	Oakland
Glen Haven	Idylwold	Wheatland
Elmhurst	Orchard Knoll	Crestland
Cedarhurst	Shady Knoll	Woodland
Willowhurst	Sunny Knoll	Meadowlands
Pinehurst	Wooded Knoll	Birdland
Oakhurst	Orchard Knob	Flowerland
Pleasanthurst	Pilot Knob	Oak Lawn
Fairhurst	Bald Knob	Maple Lawn
Maplehurst	Shady Lane	Willow lawn
Lawnhurst	Oak Lane	Cedar Lawn
Happy Hollow	Willow Lane	Shady Lawn
Shady Hollow	Sunny Lea	Lawn View
Haunted Hollow	Shady Lea	Orchard Lawn
Wood Isle	Meadow Lea	Cherry Lawn
Lake Isle	Clover Lea	Brood Lawn
Fair Isle	Highland	South Lawn
Green Isle	Homeland	Meadow Lawn
Island Home	Sweetland	Fair Lawn
Lowland	Pine Lodge	Lake View
Northland	Oak Lodge	Lawn Lake
Happy Island	Shady Lodge	Lakeholm
Lakeland	Vine Lodge	Maple Lake
Moorland	Cedar Lodge	Willow Lake
Groveland	Birchland Lodge	Island Lake
Cloverland	Mapledale Lodge	Fern Lake
Shadow Lake	Long Lake	Broad Mead
Clear Lake	Fair Lake	Fair Mead
Lake End	Shadow Mead	North Mere
Round Lake	Low Mead	Shady Nook
Crystal Lake	High Mead	Glen Nook
Lake Shore	Runnymede	West Plain
Clover Lake	Brookmede	Pleasant Plain
Loch Katrine	Grand Meadow	Plain View
Loch Haven	Meadow Lake	Oak Plain

Loch Moor	Meadow View	Maple Plain
Loch Mead	Meadow Hill	Plain Acres
Craig Loch	Meadow Home	Plainfield
Mount Royal	Meadowholm	Plaincroft
Mount Pleasant	Mead-o-Mine	Prairie Home
Mount Rose	Meadowmine	Rose Prairie
Montrose	Meadowbloom	Prairie Lawn
Fairmount	Meadowbrook	Prairie View
Mount View	Meadowdale	High Prairie
Grand Mount	Grand Mesa	Prairie Lake
Grand Mound	Mesa View	Prairie Place
Manor Hill	Mesa Place	Prairie Lane
Hillside Manor	LaBelle Mesa	Sun Prairie
Manor Place	Plain Moor	Prairie Run
Manor Lawn	Low Moor	Prairie Hill
Manor Grove	High Moor	Pride-o-the-Prairie
Clover Mead	East Moor	Shady Point
Brook Mead	Smith Moor	Orchard Point
Orchard Moor	Meadow Point	Maple Run
Moor Park	Lake Point	Birchwood Run
Broadmere	Maple Point	Oak Run
Crystal Mere	Cedar Point	Meadow Run
Low Mere	Pond Place	Low Run
Clover Mere	Meadow Pond	Rocky Run
Round Pond	Spring Run	Ridgeholm
Tanglewood Run	Springcrest	Waldkopf
Smith Run	Spring Park	Edenwald
Clover Run	Silver Spring	Lindenwald
Shadow Rill	Clean Spring	Lindenbaum
Shady Rill	Spring Brook	Edenwold
Orchard Rill	Spring Creek	Cottagewold
Pleasant Rill	Meadow Spring	Springwold
Meadow Rill	Springhurst	Cloverwold
Riverview	Fair Vale	Wild Rose
Riverside	Wood Vale	Wildwood
River Lane	Brooke Vale	Wildhelm
Rivermead	Spring Vale	Wildhurst
River Place	Oakvale	Woodfield
River Head	Lakevale	Wood Lake
River Point	Rockvale	Wood Acres
River End	Pleasant View	Wood End
River Lawn	Long View	Cottagewood
River Bend	Broad View	Plentywood
Big Bend	City View	Woodbine
Pleasant Ridge	Wood View	Cloverwood
Walnut Ridge	Hill View	Wood Lane
Maple Ridge	Grand View	Wood Croft
Oak Ridge	Spring Valley	Clearwater
Willow Ridge	Broad Valley	Sweetwater

Ridgeview
Ridge Home
Peaceful Valley
Fallawater
Gruenwald
Waterhurst
Grosserwald

Smiling Valley
Orchard Valley
Falling Water
Buena Vista
Waterview
Bruchwald
Waldhelm

Whitewater
Water's Edge
Grand Valley
Sleeping Water
Bruchwald
Watercraft

Miscellaneous Names

Of the following list of miscellaneous names, a large proportion are already in use on successful farms. Thirty-eight of them have been adopted for farms in Minnesota.

Airy Hill
Airy Knoll
Alfalfadale
Altamont
Alpine
Arden
Arrowdale
Branching Brook
Breezy Point
Brian Lake
Broadway
Brookdale
Cedercroft
Cedar Lodge
Centennial
Clearbrook
Clover Crest
Cloverdale
Clover Hill
Cold Brook
Evergreen Ridge
Excelsior
Fairfield
Fair Oaks
Fairview
Fairweather
Fanwood
Forest Hill
Lyndale
Plainview
Poplar Lane
Riverdale
Rookwood

Come Again
Corn Belt
Crossways
Cuyuna
Dairy Downs
Daisey Meadow
Double Deck
Eagle View
East View
Echo
Echo Glen
Echo Grove
Elmwood
Eureka
Edgewood
El Dorado
Elmendorf
Elmenhurst
Elmhurst
Evergreen
Ideal
Jerseyland
Lakeside
Lakeview
Lakewood
Lawnview
Lindenwood
Lucky Strike
Pinehurst
Pleasant Prairie
Red Gate
Riverside
Rose Hill

Arrowhead
Bannerland
Beechwood
Belle of Minnesota
Blue Spruce
Bonanza
Bonnie Birches
Forest Grove
Fountain Home
Gainford
Glendale
Gopher State
Grandview
Grassland
Halcyon
Hawthorn
Haycroft
Hazelnook
Hillhurst
Homestead
Orchard Hill
Osage
Our Choice
Overview
Paramount
Park Place
Perfection
Pine Grove
Pine Ridge
Pleasant Valley
Richfield
Rockwood
Seven Oaks

Shady Lane	South Shore	Spring Brook
Spring Well	Stillwater	Stony Run
Strawberry Hill	Summit	Sunnyside
Sunnyslope	Sunrise	Sunset
Sweepstakes	Sylvan Border	The Best Ever
The Knolls	Wildwood	The Willows
Willowdale	Trek End	Willow Glen
Twin Oaks	Willow Lane	Unique
Woodland	Welcome Home	Woodlawn
Westbound	Woodside	

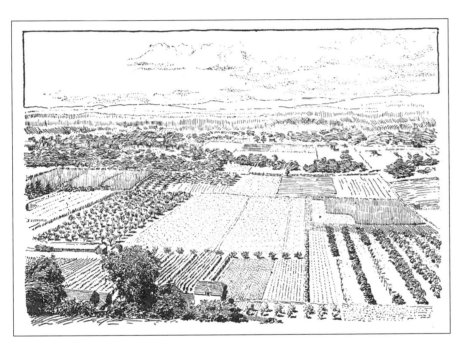

Chapter Thirteen

For The Long Haul

"What a beautiful path the cows make through the snow to the stack or to the spring under the hill! - always more or less wayward, but broad and firm, and carved and indented by a multitude of rounded hoofs." (John Burroughs)

This book has covered a lot of ground (pardon the unintended pun). Or has it? While reading through this manuscript for help in composing this closing chapter I am chagrined with each passing page to feel and see what I have left out. Hopefully this deliberately broad, almost expansive, overview I've presented has helped for that has been my sole purpose, to be helpful.

As you know I've snuck personal philosophy and politics into most of the previous chapters. With this parting shot I offer an unabashed chunk of pushy posturing as I strive to make a closing argument for farming versus agriculture, craft versus industry, fertility versus productivity, and community versus incorporation. There, I've warned you in advance. If you wish to put the book down now I'll understand and wish you well in your adventure.

In modern western societies "success" rules the day. And the contemporary take on "success" is tied directly to the rules of the franchise. If you wish to be successful by "society's" measure you must agree to a rigid franchise and work within its rules. Within these rules certain concepts or aspects are sacred - to be accepted totally and with blind faith. These concepts include;

    specialization

    competition

    ever higher profits

    ever higher technology

    ever more consumption

    convenience

    global markets

To criticize any of these concepts or aspects is considered - by government, industry, economic circles, academia, some modern religious factions, all of organized sports and most fraternal organizations - to be heresy.

If you have no problem accepting the corporate ethic you have no problem accepting all of the franchise rules. If, for you, the corporate ethic does not work you will have great difficulty with the "franchise" and the rules and making the grade to "their" notion of success.

What is this "franchise"? Why should I care? And what does it have to do with my farm dream?

It may seem like a contradiction in terms and ideas but we see clear evidence today that a selfish lifestyle leads to slavery and a selfless lifestyle leads to freedom and independence. And it all begins when we make the choice whether or not to buy into the modern socio-corporate franchise. It doesn't happen all of a sudden. You aren't asked to sign a contract. There are no initiation rites behind the clubhouse. It can be a slow process taking years or in some cases it can happen almost overnight. But no one will announce it to you. You'll have to have your eyes wide open and be looking for the signs.

It works this way. If you are greedy and selfish and looking for ease you probably have identified that you want to be rich and famous or at least rich - you want success. But what measure of success? There are many. IF you want to be successful in the eyes of most others you must accept what they consider to be evidence of success. You must work to reach the larger group's standard of success. All of that points to the status quo which today emanates from large corporations, public education, federal and state government, and modern synthetic journalism. You must accept that in order to succeed in this way you have to follow a certain path with rigid dos and don'ts. And you must pay for the opportunity to succeed by buying a franchise from everybody else. The way you buy this nebulous formless shapeless paperless thing I call a franchise (or citizenship in the socio-corporate state) is by agreeing to forfeit your independence of thought, of motivation, of allegiance, of practice, of worship, and of sacrifice. Yes, everybody has the choice; the freedom to choose what they think, why they do things, who they call their friends and community, how they do things like farming, how they worship, and what they are prepared to struggle and die for.

"Yeah, so?" you say, "Everybody knows that." Not true, for if they did they would never have bought into that franchise I spoke of.

Specifics:

If you accept the notion that the way to succeed is to specialize, do one thing only and better than anybody else, you have forfeited your right to do many and various things well and combine them into a splendid orchestration of skills and efforts the total result of which is always superior to the parts. In **agriculture** it translates to growing just tomatoes and nothing else - or - in **farming** it translates into growing tomatoes, peafowl, trout, potatoes, lamb, a lovely appreciated family, garlic, hay, pasture, wild rice, calendulas and blueberries.

If you accept the fascist notion that you are "either with us or against us" and blindly sign on to the party or group plank in order to remain a member you have forfeited your right to think and act independently and you have actually become worthLESS as a member of any group. In agriculture or agribusiness, community often gets in the way of the bottom line (profit!) because a bunch of little independent operators joined together into a community often exert themselves politically by requiring their larger neighbors to adhere to standards of conduct which benefits the community. A big factory belching toxins into the air is always going to be at odds with the small community it neighbors. The big corporate absentee -owner agribusiness outfit isn't going to care about the local church building fund, or the day care center or the corner coffee shop. And they won't contribute in any humane and immediately human sense to that community. The eccentric individual who takes it upon him or herself to be a part of the process of shaping little day to day events within the community is an essential key to the group's vitality. The child who helps at the car wash to raise money for the school, the widow woman who volunteers at the medical clinic, the farmer who donates a pig to the relief  fund, the farm laborer who donates coins to help the injured firefighter - these are community. These are individuals, who in their selfless natures, embody both a citizenship of service and the power of independence. These are individuals who have made a committment to their "place". These are folk who are in it for the long haul.

Ever higher profits and higher technologies require a mobility or transience of ethics, values, allegiance and place. When the mine has been stripped of its ore the company must move on. When a way of working can no longer be "improved" by technological advance, the technocrats deem that way of working as obsolete and they move on to new opportunities. When it suits the timber company to value the local mill town, they do. When the trees are all cut, the town and its inhabitants no longer have value to them, the company moves on. When a high tech firm needs specially educated young people for its work force, it will contibute to the local colleges so long as the curriculum suits them. When their needs are filled or their direction changes, this allegiance disappears. The socio-corporate world does not value, nor contribute to, humanity except as a market and labor force. They value profit and the means to expedite profitability, all else is expendable. This is a dead end.

Enter the small independent family farmer by the hundreds of thousands. A new land rush, a new out migration, a new wave of communities built of the immediate needs and long haul values which these farmers bring with them. Enter homemade Apple Cider at church socials and exit Diet Pepsi, enter Cotton

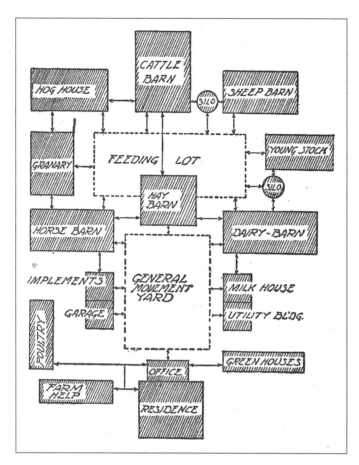

*Here is an rather clinical look at how farm buildings might be arranged to facilitate an optimum interrelationship and procedural advantage. It is doubtful that any small farm would have such a variety of buldings but with this layout you can begin to see how where buildings are in relationship to one another might make certain chores easier and even improve the overall efficiency and perhaps even **fertility** of the farm.*

and Wool and Linen and exit Dacron, enter Smoked Fish and Buttermilk and exit Cheetos and Mountain Dew, enter Corn Husking Bees and exit Video parlors, enter barn raisings and exit television, enter farm field work and exit health clubs, enter Maple Syrup and exit Kool Whip, enter homecooked meals at the big table with the family and exit McDonalds, enter young people milking cows and exit mall rats, enter small mom and pop stores in town and exit malls, enter horseback rides and exit all-terrain vehicles, enter learning and exit education, enter farming and exit agriculture, enter people and exit corporations, enter hope and exit despair. It all starts with you and your successful entry in small farming.

But its not enough just for you to get a small farm of your own. You need to be there for the long haul. We don't need a lot of folks just owning small farms. We need all of them farming and farming well.

In modern **agriculture** the working time frame is usually one year sometimes less sometimes a little more. In **farming** the working time frame should be two or three generations or more. What dictates the value of seeing farming in a longer time frame? The health and fertility of the soil, the development of improved livestock, the development of healthier plant stock, the unfolding of successful rythmns in crop rotations, the building of procedural fertility, and the continuity of family and community - all these things require many, many years. And all of these things require craftsmanship and independent thought. Each biological region has its own unique characteristics, each farm has its limitations and possiblities, each community develops its own aggragate nature. For these

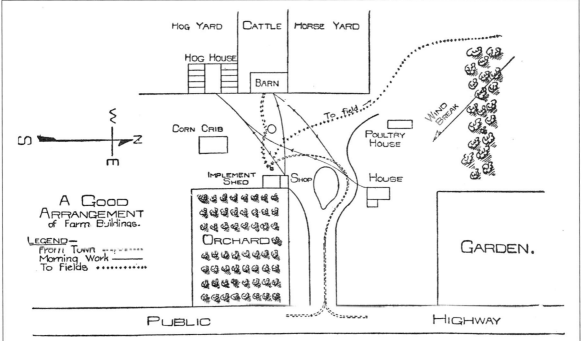

A well-arranged farmstead showing the direct and comparatively few routes that have to be taken in doing the daily work.

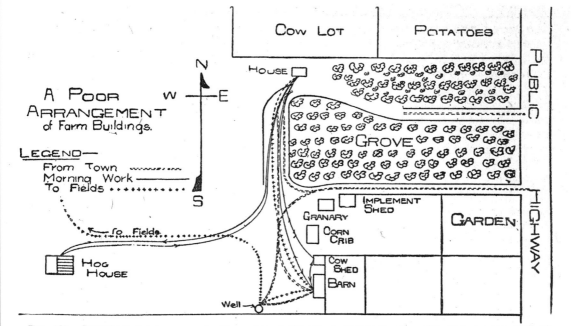

Fig. 460. Sometimes there are reasons for an apparently poor arrangement; more often they are the result of lack of thought. In either case they should be corrected as soon as possible. It is one thing to lose half an hour, unavoidably, once a month; it is another thing to lose half an hour every day the year round, especially when a little planning would prevent it.

"A fertile neighboring farm benefits you and vice versa. A thriving small town or village lends strength and depth to the farm family's living adventure. And the farmer with a fertile imagination is going to see possibilities and directions which are not visible on the direct surface."

reasons each farmer must approach his or her working world as completely unique and different from all others. There can never be a working formula for all successful and fertile small farms. Each must evolve its own. But there can be abiding values which drive the decision making process. I propose that we remove the bottom line "profit" and replace it with "fertility".

Imagine with me that every year we took a look at our balance sheet with the goal of measuring our farm's fertility - did it go up or down? If it went down, we didn't do so good and need to improve some things. If it went up, we are successful in the best way possible. Now you might say " okay, that's well and good but it doesn't pay the bills." I think you are wrong. It does pay the bills and for the long haul. Look at it this way; if your farm fertility (and general health) slips every year at some point you will have less crop to feed or sell and it will cost more to get production up. If your overall fertility factor is on the increase, it will cost you less each year to produce a saleable commodity. But the only way you get there is to fully understand fertility and work FIRST to increase it.

You might have a working understanding of <u>soil</u> fertility, as it is paramount to successful farming, but there are other aspects to a farm's fertility factor. The simple auxillary ones include reproductive fertility of livestock and plant life. This is most definitely influenced by the relative health and fertility of the soil, for it is from the soil that we receive nutrients for plants and animals which affect all aspects of their well-being. Then there are abstract notions of fertility which are equally important and certainly tied back in. One such notion is that a farm and farmer can develop and nurture a procedural fertility, which is to say that the working atmosphere and values are such that functioning aspects of the farm regularly suggest or "give birth" to new procedural models. Another way to put it is to say that the farm is open to changes in how things are done. The abstract notions also include the fertility of the neighboring countryside, the community and the farmer's imagination. A fertile neighboring farm benefits you and vice versa. A thriving small town or village lends strength and depth to the farm

family's living adventure. And the farmer with a fertile imagination is going to **see** possibilities and directions which are not visible on the direct surface.

If your soil is alive and fertile, if your family is healthy and strong, if your crops are thriving, if your livestock are fertile and healthy, if your farm is growing in possibilities and interrelationships, if your neighbor's farms are vital and healthy, if your community is happy and bustling, and if your own thoughts welcome the working suggestions which spring from this whole environment - you are successful. Notice there has been no mention of money or profits, if they're there it is as a residue to be put to some good use.

Silo filling with old-fashioned belt-drive chopper/blower.

But, yes, it is possible to speak in terms of a piece of fertile soil - and leave it at that. Though inside of this larger possibility of interconnectedness that piece of soil is only one ingredient in a mix of parts. What's in the mix is up to the participants. It can be as complex as all the previous abstract talk has suggested or it can be as direct as a mix of crops and livestock and soils with a crop rotation plan and a grazing rotation plan implemented. Coming to understand how a crop of Buckwheat, grown for weed suppression and green manure, affects the next season's grain crop and the calving ratio of the resident cow herd and thereby the total farm's fertility is arriving at craftsmanship within farming. You end up with more to sell and greater profits without that being the goal. The goal was and always should be fertility.

And it all goes back to you choosing your definition of success and allegiance. If you choose a life of service as a small farmer it may make of you an independent person who, as a good neighbor and great craftsman, enjoys a right livelihood. It doesn't get any better than that. And I have no doubt you will want to be in this for the long haul.

In parting I offer a toast:
May your poultry be as music and your grains sing songs,
may your neighbors return borrowed tools
   and your lovely daughters attract strong summer help,
may your wife enjoy watching you eat
   and may your son help with the harnessing,
may your soil speak to you in bell-like whispers
   and may your fencerows shelter pheasants,
may you never tire of looking at neat rows of kitchen
      vegetables.
may your foals know great kindness
   and may the beauty of your clover and grass bring tears
to your eyes.
   I wish for you fertility.
   Thanks for being there, LRM.

INDEX

Abandoned Properties	36
Abandoned Wells	41,47
Absentee Owners	36
Absentee Ownership	31
AC Model WC Tractor	147
Addendums	55
Advance Rumley	149
Advertised Properties	35
Agribusiness	60
Agri-Chemical Applications	41
Alfalfa	82
Allis Chalmers	148
Alternative Agriculture	87
American Cream	95
Amish Companies	98
Amish Farms	116
Amortization	53
Amortizing	27
Angus Beef Cows	80
Animal Power	91
Animal-Powered Agriculture	130
Annual Payments	31
Antifreeze	153
Apple Orchards	77
Araucana Chickens	79
Arc Welder	114
Are Horses Coming Back?	136
Asking Price	43
Assumable	43
Assumable Balance	43
Attorney	44,53
Auctioneer	128
Auctions	96,126
Auxillary Motors	153
B Tractor	146
Back Taxes	44,57
Bags	124
Balloon	52
Banamine Paste	137
Banker	65
Banking Fraud	53
Banks	36
Barley	80
Barn	156
Barns	154
Barrel Pump	149
Beef	82
Belt Drive Wheel	153
Berries	82
Bidding	129
Binder	92
Biodynamic	41,77
Biological Diversity	87
Biothane or Nylon Harness	97,137
Black East Indian Ducks	79
Blue Andalusian	79
Borrowed the Money	65
Borrowing	51,52
Bracing End Posts	172
Brahms	89
Breaches	55
Broiler Chickens	80
Broiler Huts	81
Broke	94
Brush	137
Buckrake	125
Buckwheat	77
Butte Tablets	137
Budgets	126
Buff and Penciled Runner Ducks	79
Buff Cochin	79
Buffer Ag Lands	46
Building Fence	172
Building Plans	163
Building Zones	46
Buildings	41,141
Buying on contract	27
Buzz Saws	153
Cabbage	77
Calves	78
Canning Facilities	119
Cape Feathers	79
Case	148
Cash Cropping	77
Cattle Barn	157
Cheapest Land	26
Cheese	82
Cheese-making	119
Chemical Applications	46
Chemically Farmed	41
Chest Freezer	119
Chicks	82
Churches	40,46
Cider Mill	77
Cider Press	119
Clear Title	55
Clevises	138
Closing	56
Closing Costs	56
Clover Grass	77
Collars	96,136
Collateral Pledges	39
Combine	92
Common Root-Words and Suffixes 176	
Composting Facilities	155
Conestoga	98
Conformation	131
Contagions	41
Contagious Livestock Diseases	47
Contingency	55
Contract Collection	56
Contract to Purchase	53
Contracts	56
Convenience	63
Corn	82
Corn/Bean Planter	100
Corn Crib	155,166,167
Corn Harvest	81
Costs	94
Cottage Industries	125
Counsel	62
Counter Offer	55
County Offices	47
Cow Stall	159
Cows	78
Crop Rotation	77
Crop Value/Income	28
Crops	47,76,80
Cultivator	101
Curry Comb	137
Custom	92
Custom Meat Cutter	81
Dairy Cattle Barn	159
Dairy Heifers	82
Dairy Steers	82
Deed of Trust	39
Deed Restrictions	44
Designing Your Farm	67
Development	30
Diesel	153
Direct Marketing	76,82
Disc	99
Dislikes	48
Diversity	68
Dollars Per Acre	26
Dominque	79
Double Trees	97,137
Down Payment	54
Draft Horses	80,131
Draw-Type Tractor Implements	138
Drawbacks to Using Horses	135
Drill	106
Drill Press	114
Dumps	41,47
Earnest Money	54
Earnest Money Agreement	38,53,54,55
Easements	39,44
Economic Balance	28
Eggs	79,82
Eggs/Dairy	82
Electrical	40
Enclosures	55
Enterprise Data	59
Environmental Concern	45
Equipment	97
Escrow	54,56
Escrow Closing	39
Escrow Company	56
Ethics	185
Eveners	97
Exercise Pens	168
Expectations	48
Explosive Hazard	153
Extension Agents	47,59
Ewes	81
Facilities	138
Family	84
Family's List	85

Farm Auction	128	Good Realtor	38,42	Internal Combustion	91,153
Farm Community	59	Good Arrangement of Farm Buildings		Irrigation Equipment	39
Farm Economists	59		187	Jersey Cows	78
Farm Implement Stores	125	Goose	82	John Deere	146,148,149
Farm Implements	119,121	Gould	89	Juice	78
Farm Layout	141	Grab Hook	97	Jurisdictional Districts	45
Farm Lenders	23	Grain Binder	103	Lambing Barn	162
Farm Machinery	132	Grain Drill	101	Land Sale Contract	39,53
Farm Part-Time	66	Grain Harvest	92	Land Sales Agreement	31
Farm Repairs	113	Grain Storage	155	Land Use	45
Farm Sale	127	Grains	77,81	Landlocked	44
Farm Shop, The	113	Grazing	81	Lawyer	38
Farm Supply Stores	125	Green Houses	155	Layers	82
Farm Wagon	98	Green Manure Crops	77	Layout	186
Farm's Fertility Factor	189	Grooming	137	Leather Harness	137
Farmall M	146	Gross Return	28	Legal Council	43
Farming History	47	Ground-Drive	120,145,153	Legal Trouble	55
Farrowing Shed	157	Guarantees	39	Legume Mix	82
Feather Plate	160,161	Guarantor	39	Liens	30,39,44
Federal Farm Credit	44	Halflingers	95	Likes	48
Federally Guaranteed Loan Programs		Hammermills	153	Line Shaft	116
	53	Hand-Labor	91	Listing Agreement	37
Feeder Hogs	80	Harmonic Balance	87	Livestock	78,80
Feeding Barn	168	Harness	96,132,136	Livestock Disease	41
Fence Boards	124	Hay	79	Livestock Feeds	124
Fence Posts	124	Hay Loader	102,106	Living Expenses	60
Fences	164,172	Health Food	79	Loafing Shed	154
Fertile Eggs	79	Healthy Farm	90	Long Term	52
Fertility	186	Heifers	80	Lubricants	124
Fertilizers	124	Hitch Gear	132,137	Lumber	124
Fewer Acres	33	Hog House	162,163	Manure Pit	157
Field Hitches	138	Hogs	81	Manure Spreader	97
Field Partition Plan	141	Holding Companies	37	Manure Storage	46
Financing	40,51,52	Hoof Care Tools	137	Maple Syrup	80,81
Fine Furniture	78	Housing of Workhorses	142	Mapping	69
Fire Coverage	40	Horowitz	89	Market Garden	28, 29,80,81
Fire District	40	Horse-Logging	135	Market Identification	175
Fire Insurance	40	Horsepower	118	Markets	33,40,46
Fire Insurance Coverage	46	Horse-Powered Farm	91	Massey Harris	148
Fire Protection	40	Horse Stall	139	McCormick Deering Model 06	150
Fires	153	How Many Acres	135	Meat	82
Fish	77	How Many Horses	94,131	Metal Stanchion	159
Fixed	52	How Much Does a Work Horse Eat?		Metal Working	114
Flatbed	98		132	Milk Cows	82
Fly-tying	79	How Much Should I Pay	24	Milk House	171
FmHA	53	How to Buy Tools & Implements	125	Milk Production	79
Food Processing	119	How to Know What You'll Need	110	Milking Parlour	154,155
Ford	146,148	Hydraulics	120,145,153	Mille Fleur	79
Ford Series	146	Implement Shed	170	Mineral	39
Forecart	103,137,138,153	Improvements	32,39	Miscellaneous Names	180
Foreclosure	43,44	Incubation Room	162	Mixed Crop and Livestock	17,80
Forge	114	Indebtedness	31	Mixed Power	153
Four Abreast Evener	97	Insurance	56	Mixed Power Farming	152
Fuel Storage	149	Insurance Company	43,53	Mixed Woodlot	77
Fuel Storage Tank	151	Intensive High Density Grazing	77	Model B Allis Chalmers	144
Fuels	124,153	Intensive Perennial Cropping	28	Model C Tractor	147
Fraternal Organizations	40	Intensive Rotational Grazing	78	Money	51
Frugality	126	Interchangeability	152	Money Transfer	53
G. I. Loans	53	Interconnectedness	190	Monocultural Operations	112
Gas	30,153	Interest Points and Fees	52	Monthly Payments	31
Gates	174	Interest Rate	31,52	Mortgage	39,52
Geese	82	Intermediate Technology	77	Mortgage Payments	31
Get Started	134				

Motor Oils	153	Potato Planter	107	Sales Data Files	29
Mower	102	Potatoes	81	Saw Mills	153
Mule	92,132	Potting Shed	162	School	39,40,46
Musical Instruments	78	Poultry	79	Security	39
Naming Your Farm or Ranch	175	Poultry Facilities	155	Security Interest	40
Neckyokes	97,137	Poultry House	157	Seed	124
Negotiations	56	Power	46	Seed Innoculants	124
Net Income	28	Power-Take-Off	120,153	Seller Financing	53
New Leather Harness	97	Preliminary Title Report	56	Selling Privately	36
Nighttime Field Work	46	Price per Acre	28	Shackles	138
Nitrogen Fixing Legume	77	Prices	104	Sheds	141,154
No Buildings	154	Principle Payment	31	Shed Space	155
Non-Compliance	55	Private Sale	129	Sheep	77
Non-Farm	66	Processing Kitchen	155	Sheep Barn	168
Norwegian Fjords	95	Producing Mares	133	Shopping List	126
Number of Rods of Fence Required	173	Production Credit	44	Silage Bunker	154
		Production Levels	39	Side-Delivery Rake	102
Nurse Crop	81	Property Taxes	27,29,39,45	Silo Fillers	153
Oat Grain Bundles	105	Profitability	185	Silver Laced Wyandotte	79
Oats	80,82	PTO	145	Single Tree	97,137
Offer	55	Pull-Type	120,145,153	Single Work Horse	79
Offer to Purchase	54	Pull-Type Implements	152	Size	111
Offering Price	31	Pullets	82	Small Down Payment	31
Oil	30	Pumpkins	81	Smoke House	155
Older Tractors	123,146	Purchase Contract	55	Social Services	33
Oliver 550	150	Purebred Draft Horses	133	Soil Amendments	124
One-room Milkhouse	171	Purebred Mare	96	Soil Biology	123
Operating Expenses	65	Purebred Trios	79	Soil Tests	47
Orchard	41,47,80	Quarterly Payments	31	Soil Types	41
Ordinance	39	Qualified Buyer	42	Soils	47
Organic	41	Rambouillet Ewes	80	Solvents	153
Organic Cereals and Flours	77	Raw Land	154	Specialized Implements	105,122
Organic Formula	123	Raw Milk	82	Speckled Hamburg	79
Organic Orchard	77	Real Estate Agencies	36	Spend for Horses?	132
Optimum Interrelationship	186	Real Estate Contract Law	55	Spike Tooth Harrow	99
Outbuildings	39,154	Realtors	33,37,53,55	Spreaders	98
Outside Job	66	Receipts	32	Spring-Tooth Harrows	100
Oxyacetylene Rig	114	Recent Sales Data	30	Stabling	153
Owed	39	Rectal Thermometer	137	Stallion and Mare Barn	169
Owner Financing	38	Refrigeration	40	Start a Horse at Hard Work?	133
Packing Sheds	77,155	Refrigeration Unit	155	Stationary Power Supply	116,153
Paddocks	81	Relevant Districts	39	Steers	80
Partridge Cochin	79	Religion	40	Storage Tanks	39
Payment	52	Replacement Heifers	82	Straddle Row	101
Peas	77	Restaurant Broilers	79	Strawberries	82
Percentage Down Payment	38	Retail	126	Structural Condition of Buildings	47
Perimeter Fencing	165	Return per Acre	28	Subsurface Rights	44
Permanent Pasture	78	Return on Investment	28	Suburban	46
Personal Philosophy	182	Riding Plow	98	Success	183
Pest Remedies	124	Right to Farm Laws	46	Suffolks	95
Pheasants	79	Roller	100	Suggestive Suffixes and Root Words	176
Physical Hazards	41	Roller-packers	100		
Place Names	176	Root Cellars	119	Supplies	123
Poison-Free	41	Root Families	77	Surplus Grains	81
Pond Soil	77	Rotational Grazing	82	Sustainability	87
Ponds	41,47,80	Rotate Pasture	81	Swivel Shackle Grab Hook	97
Poor Arrangement of Farm Buildings	187	Rotten Harness	97	System Design	48
		Royal Palm Turkeys	79	System Examples	75
Pork	82	Running Gear	98	System Possibilities	67
Possibilities	190	Rye Grass	80	Tack	136
Post and Beam	78	Saddle Horses	80,131	Tax Assessor	29
Potato Digger	120	Safely Store Feed	140	Tax Lot Plot Maps	30

Tedding	102	Types of Farm Gate	174	Well-Trained Team	96
Third Party	56	Unadulterated Meats	81	Wells	40
Third Party Financing	40	Unencumbered Property	57	What	17
Third Property	52	USDA	111	What Kind of Farming	17
Three Point	120	Used Collars	97	What Sort of Buildings	155
Three Point Hitch	152	Used Harness	97	What Sort of Farm	67
Thresher	103,153	Utilities	40,46	What Tools	109
Threshing Machine	92	Vacation	62	Wheat	80
Thrift	126	Values	83,185	Wheeler	97
Tie-Stall	138	Variable	52	Where	18,35
Timber Rights	39	Vegetable Processing	118	Why	14
Title Company	44	Veterinarian Supplies	124	Wildlife Shelters	141
Title Insurance	39,56	Veterinarians	47,137	Windbreaks	141
Title Search	37,39,55,56	Vice	114	Winter Holiday	79
Toast	191	Vintage Apple	78	Wire	124
Tool Shed	162	W-6	151	Without a Realtor	42
Tools	108,114	WC Allis Chalmers	145	Without Electricity	114
Tractor Shopping	146	Walking Plow	98	Woodlots	41,47,82
Tractor-Powered Farm	143	Water	46	Work Horse Barn	168
Trade	52	Water Purity	40	Work Horses	82
Transfers	56	Water Rights	45	Work Horses or Mules	130
Transplanter	107	Water Rights Recorded	39	Working Horses	91
Treadmill	118	Water Wheel	118	Working Life of a Horse	132
Tree Farmer	78	Watering Tanks	124	Wool	81
Triangular Enclosures	172	Watermelons	81	Wrenches	114
Triple Tree	97	WD-6	151	Zoning Changes	30
Twine	124	Weed Suppression	81	Zoning Restrictions	39
Two-Room Milkhouse	171	Well Arranged Farmstead	187		

Photo credits: Elizabeth Buchser (page) 14. Judith Hoffman 122, 123, 124, 130, 131, 132, 133, 135. Kristi Gilman-Miller 19, 48, 54, 57, 59, 61, 63, 79, 80, 81, 82, 89, 93, 95, 109, 110, 111,112, 121, 124, 125, 127, 128, 134, 138. Juliet Miller 91. Lynn Miller 16, 78. Nancy Roberts 182. Other materials from SFJ archives.

About the author: Lynn Ralph Miller was born in Kansas City in 1947. Most of his youth was spent in So. Calif. suburbs. In 1965 he went to college in San Francisco. In 1969 he moved to Oregon and has been farming, painting, and writing since. In 1976 he created 'Small Farmer's Journal' for which he has functioned as editor and publisher. He and his wife and youngest daughter live and work on a remote eastern Oregon ranch.

Suggested Additional Reading

Small Farmer's Journal

Featuring Practical Horsefarming

An international quarterly fiercely supportive of independent diversified small-scale family based agriculture. This large attractive publication includes outstanding articles on any and every subject of interest to folks dedicated to a healthful lifestyle on a successful small farm.

Subscribe today with NO risk. If you aren't completely satisfied we will refund all your money. Send $24 [U.S.] or $31 [U.S. funds, Canada and other countries] for a one year (4 issues) subscription to;

Small Farmer's Journal

Dept. DBFS, P.O. Box 1627, Sisters, Oregon 97759
or telephone **541-549-2064** [9-5 PST] for credit card orders

Farmer's Book Service

Small Farmer's Journal is pleased to be able to offer an ever changing list of books of guaranteed honest value. Through our **Farmer's Book Service** we strive to offer a selection of appropriate book titles which you may have difficulty finding in your book store. These books cover subjects as far flung as Mushrooms, Poultry, Gardening, Carriage Restoration, Oxen, Recipes, Weeds, Barns, Marketing Farm Produce, Goats, Root Cellaring, Sheepdogs, and so much more. For a complete up-to-date listing of available titles write to

Small Farmer's Journal
PO Box 1627, Sisters, Oregon 97759 or phone
541-549-2064

TRAINING WORKHORSES / TRAINING TEAMSTERS

A NEW text combining two books in one and including 482 photographs and hundreds of drawings on 352 pages. This text covers the subjects of;
*training horses to work in harness
(on the farm, in the woods and on the road)
correcting behavior problems with work horses
and training people to drive and work horses.*

Soft cover $24.95 (includes Postage and Handling)
Hard Cover $43.95 (includes P & H)

Please send check or money order to
SFJ, Dept DBFS, P.O. Box 1627, Sisters, Oregon 97759
M/C, VISA or Discover orders accepted by phone at (541)549-2064

Your complete satisfaction is guaranteed. If this book does not meet your expectations return it for a full refund.

WORK HORSE HANDBOOK

It has become a classic and **the** standard reference. This popular, highly regarded text is filled with current information and hundreds of photographs and drawings. It is a sensitive and intelligent examination of the craft of the teamster.

From care and feeding thru hitching and driving; every aspect is covered. Find out for yourself why this book is considered by thousands of people to be THE volume on working horses in harness.

$16.45 each for **The Work Horse Handbook**
Send check or money order:

SFJ, Dept DBFS, P.O. Box 1627, Sisters, Oregon 97759
M/C, VISA or Discover orders accepted by phone at (541)549-2064

Your complete satisfaction is guaranteed. If this book does not meet your expectations return it for a full refund.

TEN ACRES ENOUGH
The Small Farm Dream is Possible

Edited by Lynn R. and Ralph C. Miller

Two books in one!

"TEN ACRES ENOUGH" is a complete, unaltered reprint of the 1864 Classic.

"THE SMALL FARM DREAM IS POSSIBLE" are essays by Ralph and Lynn Miller which bring the original classic up to date.

This work includes excellent information on the cultivation of small fruits and orchards. It also provides insights into the management of intensive diversity on a small farm which includes market gardening, forage and livestock.

To get your copy of the book, send $13.45 for the softcover or $21.45 for the hardcover edition.
(Includes postage and handling)

Please send check or money order to
SFJ, Dept DBFS, P.O. Box 1627, Sisters, Oregon 97759
M/C, VISA or Discover orders accepted by phone at
(541)549-2064

Your complete satisfaction is guaranteed. If this book does not meet your expectations return it for a full refund.

Why Farm

*Selected Essays & Editorials
By Lynn R. Miller*

In 1976 Lynn R. Miller conceived of the idea of **Small Farmer's Journal** and went to work as it's part-time editor (until a "legitimate professional" could be found). Twenty two years later he's still in the editor's box writing essays and editorials, both scathing and pensive, which have, in their raw story telling form, met with warm approval by thousands of SFJ readers. In this new book Miller has updated and reworked a compilation of his early editorials and essays. This may be an important book on modern agrarian actualities and possibilities by a man who is fond of being called a "muckraker" for all of its fertile connotations.

94 pages. Soft cover. $11
(includes postage & handling)

Please send check or money order to
SFJ, Dept DBFS, P.O. Box 1627, Sisters, Oregon 97759
M/C, VISA or Discover orders accepted by phone at (541)549-2064

Your complete satisfaction is guaranteed. If this book does not meet your expectations return it for a full refund.